International Materials Research Meeting in the Greater Region 2017

Current Trends in the Characterisation of Materials and Surface Modification

IOP Conference Series: Materials Science and Engineering
Volume 258

Saarbrucken, Germany
6 – 7 April 2017

ISBN: 978-1-5108-5060-6
ISSN: 1757-8981

Printed from e-media with permission by:

Curran Associates, Inc.
57 Morehouse Lane
Red Hook, NY 12571

Some format issues inherent in the e-media version may also appear in this print version.

Printed by Curran Associates, Inc. (2017)

For permission requests, please contact the Institute of Physics
at the address below.

Institute of Physics
Dirac House, Temple Back
Bristol BS1 6BE UK

Phone: 44 1 17 929 7481
Fax: 44 1 17 920 0979

techtracking@iop.org

Additional copies of this publication are available from:

Curran Associates, Inc.
57 Morehouse Lane
Red Hook, NY 12571 USA
Phone: 845-758-0400
Fax: 845-758-2633
Email: curran@proceedings.com
Web: www.proceedings.com

Table of contents

Volume 258

International Materials Research Meeting in the Greater Region: "Current Trends in the Characterisation of Materials and Surface Modification"

6–7 April 2017, Saarland University, Saarbrücken, Germany

Published online: 26 October 2017

Preface

Papers

IOP Conf. Series: Materials Science and Engineering **258** (2017) 011001 doi:10.1088/1757-899X/258/1/011001

UNIVERSITÄT DES SAARLANDES

EEIGM

INTERNATIONAL MATERIALS RESEARCH MEETING IN THE GREATER REGION

Current trends in the characterization of materials and surface modification

6-7 April, 2017

Saarland University, 66123 Saarbrücken

E·U·S·M·A·T

IOP Conf. Series: Materials Science and Engineering **258** (2017) 011001 doi:10.1088/1757-899X/258/1/011001

1. Preface

Dear ladies and gentlemen,

On 6th and 7th of April 2017 took place the "International Materials Research Meeting in the Greater Region" at the Saarland University, Saarbrücken, Germany. This meeting corresponded to the 9th EEIGM International Conference on Advanced Materials Research and it was intended as a meeting place for researchers of the Greater Region as well as their partners of the different cooperation activities, like the EEIGM program, the 'Erasmus Mundus' Advanced Materials Science and Engineering Master program (AMASE), the 'Erasmus Mundus' Doctoral Program in Materials Science and Engineering (DocMASE) and the CREATe-Network.

On this meeting, 72 participants from 15 countries and 24 institutions discussed and exchanged ideas on the latest trends in the characterization of materials and surface modifications. Different aspects of the material research of metals, ceramics, polymers and biomaterials were presented. As a conclusion of the meeting, the new astronaut of the European Space Agency Dr. Matthias Maurer, who is an alumni of the Saarland University and the EEIGM, held an exciting presentation about his activities.

Following the publication of selected papers of the 2009 meeting in Volume 5 and 2012 meeting in Volume 31 of this journal, it is a great pleasure to present this selection of 9 articles to the readers of the IOP Conference Series: Materials Science and Engineering. The editors are thankful to all of the reviewers for reviewing the papers. Special praise is also given to the sponsors of the conference: European Commission within the program Erasmus Mundus (AMASE and DocMASE), Erasmus+ (AMASE), and Horizon2020 (CREATe-Network, Grant agreement No 644013): the DAAD (Alumni Program), and the German-French University (PhD-Track).

All papers published in this volume of *IOP Conference Series: Materials Science and Engineering* have been peer reviewed through processes administered by the Editors. Reviews were conducted by expert referees to the professional and scientific standards excepted of a proceedings journal published by IOP Publishing.

Frank Mücklich Dirk Bähre

IOP Conf. Series: Materials Science and Engineering **258** (2017) 011001 doi:10.1088/1757-899X/258/1/011001

2. Conference topics

➢ **Current trends of materials characterization**

- Microscopy
- Diffraction
- Spectroscopy
- Different techniques for characterization of material properties

➢ **Surface modification**

- Coatings
- Structuring techniques
- Surface functionalities
- Surface properties

IOP Conf. Series: Materials Science and Engineering **258** (2017) 011001 doi:10.1088/1757-899X/258/1/011001

3. Organization

Scientific Committee and Editors:

Frank Mücklich, Dirk Bähre, Magnus Odén, Janis Varna, Zoubir Ayadi, Marc Anglada, Jean-Françoise Pierson, David Horwat

Conference Organization:

Flavio Soldera, Nataliya Lyubenova, Nadine Kreutz, Federico Lasserre, Claudia Heß, Adrian Krüger

E·U·S·M·A·T

EEIGM

UNIVERSITÄT DES SAARLANDES

UNIVERSITÉ DE LORRAINE

UNIVERSITAT POLITÈCNICA DE CATALUNYA BARCELONATECH

LULEÅ UNIVERSITY OF TECHNOLOGY

Linköping University

Programs:

amase master

DocMASE

EEIGM

EEIGM

Sponsors:

Funded by the Erasmus+ Programme of the European Union

DAAD

Université franco-allemande Deutsch-Französische Hochschule

European Union's Horizon 2020 - Grant agreement No 644013.

IOP Conf. Series: Materials Science and Engineering **258** (2017) 011001 doi:10.1088/1757-899X/258/1/011001

4. Conference impressions

IOP Conf. Series: Materials Science and Engineering 258 (2017) 011001 doi:10.1088/1757-899X/258/1/011001

5. List of the participants

No	Surname	Name	Institution	Country
1	Acevedo	Diego	Universidad Nacional de Río Cuarto	Argentina
2	Akhtar	Farid	Luleå University of Technology	Sweden
3	Alnjiman	Fahad	University of Lorraine	France
4	Alvi	Sajid	Luleå University of Technology	Sweden
5	Andre	Stephane	University of Lorraine	France
6	Anglada	Marc	Polytechnic University of Catalonia	Spain
7	Aristizabal	Katherine	Saarland University	Germany
8	Arya	Visheet	Saarland University	Germany
9	Ayadi	Zoubir	University of Lorraine	France
10	Bähre	Dirk	Saarland University	Germany
11	Balach	Juan	Universidad Nacional de Río Cuarto	Argentina
12	Barrirero	Jenifer	Saarland University	Germany
13	Barth	Danielle	University of Lorraine	France
14	Borroto Ramirez	Alejandro	University of Lorraine	France
15	Britz	Dominik	Saarland University	Germany
16	Broering Chaar	Ana	Linköping University	Sweden
17	Chen	Yu-Hsiang	Linköping University	Sweden
18	Choquet	Patrick	Luxemburg Institute of Science and Technology	Luxemburg
19	Czerwiec	Thierry	University of Lorraine	France
20	De Melo Sanchez	Claudia	University of Lorraine	France
21	Di Stasio	Luca	University of Lorraine	France
22	El Azhari	Idriss	Saarland University	Germany
23	Fang	Shiqi	Polytechnic University of Catalonia	Spain
24	Forouzan	Farnoosh	Luleå University of Technology	Sweden
25	Funk	Frank	Saarland University	Germany
26	Gachot	Carsten	Technical University of Vienna	Austria
27	Gaddam	Raghuveer	AB Sandvik Materials Technology	Sweden
28	Giba	Alaa	University of Lorraine	France
29	Hauet	Thomas	University of Lorraine	France
30	Hemmer	Eva	University of Ottawa	Canada
31	Horwat	David	University of Lorraine	France
32	Hsu	Chia-Hui	Saarland University	Germany
33	Ismail	Mostafa	University of Oulu	Finland
34	Jamart	Brigitte	University of Lorraine - EEIGM	France
35	Kallel	Ali	University of Sfax	Tunisia
36	Kharkhan	Daria	University of Lorraine	France
37	Kheradmand	Nousha	Norwegian University of Science and Technology	Norway

IOP Conf. Series: Materials Science and Engineering **258** (2017) 011001 doi:10.1088/1757-899X/258/1/011001

38	Khurana	Kanupriya	Polytechnic University of Catalonia	Spain
39	Kreutz	Nadine	Saarland University	Germany
40	Krüger	Adrian	Saarland University	Germany
41	Leber	Andreas	Massachusetts Institute of Technology	USA
42	Lee	Juseok	Korea Institut of Science and Technology	South Korea
43	Leibenguth	Peter	Plansee SE	Austria
44	Lennart	Wallstrom	Luleå University of Technology	Sweden
45	Luo	Jiaqi	Saarland University	Germany
46	Lyubenova	Nataliya	Saarland University	Germany
47	Mateo	Antonio	Polytechnic University of Catalonia	Spain
48	Maurer	Matthias	European Space Agency	Germany
49	Mücklich	Frank	Saarland University	Germany
50	Nakamura	Leandro	Saarland University	Germany
51	Neikter	Magnus	Luleå University of Technology	Sweden
52	Nikjoo	Dariush	Luleå University of Technology	Sweden
53	Olguin	Andres	Saarland University	Germany
54	Öztürk	Utkundeniz	Polytechnic University of Catalonia	Spain
55	Pauly	Christoph	Saarland University	Germany
56	Pilloud	David	University of Lorraine	France
57	Pineda Huitron	Rosa Maria	Luleå University of Technology	Sweden
58	Ramos Moore	Esteban	Pontificia Universidad Católica de Chile	Chile
59	Rittgen	Kai	Institute for New Materials	Germany
60	Roa Rovira	Joan	Polytechnic University of Catalonia	Spain
61	Roland	Michael	Saarland University	Germany
62	Sandfield	Anna Lyn	Oregon State University	USA
63	Shing	Shika	Luleå University of Technology	Sweden
64	Soldera	Flavio	Saarland University	Germany
65	Taher	Mamoun	ABB Corporate Research Center and Ångström Laboratory	Sweden
66	Thielen	Mathias	Saarland University	Germany
67	Vitzthum	Valerie	University of Lorraine	France
68	Vuorinen	Esa	Luleå University of Technology	Sweden
69	Wang	Yong	Southwest University of Science and Technology	China
70	Weidner	Maren	Saarland University	Germany
71	Zhang	Hanzhu	Luleå University of Technology	Sweden

IOP Conf. Series: Materials Science and Engineering **258** (2017) 011002 doi:10.1088/1757-899X/258/1/011002

Peer review statement

All papers published in this volume of *IOP Conference Series: Materials Science and Engineering* have been peer reviewed through processes administered by the proceedings Editors. Reviews were conducted by expert referees to the professional and scientific standards expected of a proceedings journal published by IOP Publishing.

IOP Conf. Series: Materials Science and Engineering **258** (2017) 012001 doi:10.1088/1757-899X/258/1/012001

Correlation between the dielectric and the mechanical behavior of cellulose nanocomposites extracted from the rachis of the date palm tree

A Ladhar[1], M Arous[1], H Kaddami[2], Z Ayadi[3*], A Kallel[1*]

[1] LaMaCoP, Faculty of Sciences of Sfax, University of Sfax, BP 3018 Sfax, Tunisia
[2] LCBM, Faculty of Sciences and Technologies, Cadi-Ayyad University, 40000 Marrakech, Morocco
[3] EEIGM, Equipe PMP-304, Nancy, France

*zoubir.ayadi@univ-lorraine.fr; ali.kallel@fss.rnu.tn

Abstract. In the present study, the dielectric and mechanical properties of natural rubber (NR) based nanocomposites are investigated. Cellulose nanofillers are used in two forms as reinforcing phase: nanofibrillated cellulose (NFC) and cellulose nanocrystals (CNC). In the dielectric study, different relaxation phenomena are detected: the α dipolar relaxation, the lignin and hemicelluloses relaxation, the water dipoles relaxation, the interfacial polarization and the ionic conduction. For the interfacial polarization, the dielectric strength $\Delta\varepsilon$ showed lower values for NFC-filled nanocomposites than CNC-filled samples. It was explained with higher interactions between induced dipoles and lower mobility, assuring a better adhesion between the NR and the NFC. Moreover, in tensile tests, the elastic modulus increases with filling indicating the reinforcement effect of nanofillers. In addition, the NR-NFC nanocomposites display the highest tensile modulus. This result shows the higher compatibility of NFC with the NR matrix, and the ensuing higher filler/matrix adhesion. In dynamic mechanical analysis (DMA), a significant reinforcing effect of NFC was shown. This effect is manifested with the high storage modulus E', suggesting that the interactions between the NR matrix and the NFC fibers were stronger.

1. Introduction

Nowadays, two major problems are encountered in the polymer industry: (*i*) the problem of pollution due to the use of materials of petroleum origin (*ii*) the properties limit of some used materials. Therefore, production of innovative bionanocomposites derived from natural sources is currently one of the main points of interest in the academic and industrial areas of material research. Studies based on bio-nanocomposites using cellulose as reinforcing phase have been reported by many research groups in recent years [1, 2]. Cellulose is the most abundant polysaccharide on earth. It is a highly ordered polymer of cellobiose (D-glucopyranosyl-β-1,4-D-glucopyranose) chains aggregated by numerous strong intermolecular hydrogen bonds between hydroxyl groups of adjacent macromolecules. Cellulosic nanofibers present crystalline domains separated by less ordered ones called amorphous regions. Crystalline and amorphous domains are found in native cellulose fibers in variable ratios, as a function of the plant species, with the growing conditions or the part of the plant.

In literature, cellulose nanofillers were used as reinforcing phase on polymers based nanocomposites [2, 3]. When small amounts of cellulose nanofillers are introduced into the matrix, some changes in Young's modulus, tensile strength, thermal stability, dielectric strength and electrical conductivity had been brought. The current study is in keeping with the general pattern aiming at the valorization of nanocellulose from the rachis of the date palm tree. Cellulose nanocrystals (CNC) and nanofibrils (NFC) are used as reinforcing phase for the Natural Rubber (NR) polymer. Natural rubber (NR) has a low glass transition temperature, T_g, soft elastomer characteristics at room temperature, and good elastic and adhesive properties [4]. The choice of the matrix was dictated by the fact that it is a natural polymer, often reinforced with nanoparticles and available as latex. Natural nanocomposites can be used in automotive industry, biomedical applications, packaging industry [5], humidity sensors [6] and microporous separators in batteries [7].

This work aims to study the effect of the addition of cellulose nanofillers on the mechanical and the dielectric properties of NR-based nanocomposites.

2. Experimental

2.1. Materials

2.1.1. Polymer matrix. Natural rubber (NR) was kindly supplied as NR latex by Michelin (Clermont-Ferrand, France) and used as matrix material. It contained spherical particles with an average diameter around 1 μm and its solid content was about 50 wt%. The density of dry NR, ρ_{NR}, was 1g.cm^{-3} and it contained more than 98% of cis-1, 4-polyisoprene [8]. Its glass transition temperature was around -61 °C.

2.1.2. Nanofibrillated cellulose and cellulose nanocrystals. (NFC) and (CNC) were extracted from the rachis of the date palm tree. Colloidal suspensions of NFC and CNC in water were prepared as described elsewhere [9, 10]. The rachis of date palm tree was mashed and extracted using soxhlet with a toluene/ethanol mixture (38/62 v/v) to remove pigments, lipids and waxes. The obtained cell walls were then dried at 80 °C for 2 hours. After wet grinding in a blender, the fibers underwent an alkaline treatment. Then, the fibers were bleached with sodium chlorite to remove lignin. The main difference between the preparation of NFC and CNC was the last step in which the bleached cellulose was disintegrated by pumping through a microfluidizer processor (Model M-110 EH-30) instead of submitting it to an acid hydrolysis treatment. The solid content of the suspensions was around 0.3 wt%. (NFC) had an average length of 700 nm ~ 1100 nm and an average width of 8 nm ~ 12.5 nm while (CNC) had an average length of 100 nm ~ 400 nm and an average width less than 10 nm [8]. In this study, a series of nine samples was used (Table1). The numbers indicate the nanoparticle content, in weight.

Table 1. Codification and composition of nanocomposite films based on NR.

Samples		Nanofiller content (%)
NR		0
NR-NFC1	NR-CNC1	1
NR-NFC2.5	NR-CNC2.5	2.5
NR-NFC7.5	NR-CNC7.5	7.5
NR-NFC15	NR-CNC15	15

IOP Conf. Series: Materials Science and Engineering **258** (2017) 012001 doi:10.1088/1757-899X/258/1/012001

2.2. Experimental methods

2.2.1. Dielectric measurements. Dielectric measurements were carried out using a Novocontrol Impedance Spectrometer based on an Alpha analyzer and a Quatro temperature controller in the frequency range 0.1 Hz – 1 MHz and in the temperature range from -100 °C to 200 °C in steps of 10 °C and at an oscillation voltage of 1 V. A film of each sample was placed between two brass plated electrodes (diameter 20 mm) of a parallel plate capacitor. The sample was mounted in a cryostat, and the temperature was controlled by a heated gas stream of nitrogen, evaporated from the liquid state. The temperature was controlled by a nitrogen jet (Quatro, Novocontrol), with an uncertainty of 0.1 °C during every frequency sweep.

According to the planar capacitor rule, the complex dielectric function is expressed as [11]:

$$\varepsilon^*(\omega) = \varepsilon'(\omega) - j\varepsilon''(\omega) \qquad (1)$$

The accuracy of Alpha impedance measurement is 0.01%.

2.2.2. Tensile tests. The mechanical strength of various films was measured with a RSA3 (TA Instruments, USA) with a load cell of 100 N to identify any change in their physical properties upon the addition of NFC or CNC into the NR film. The specimens were prepared to be 5 mm in width and 40 mm in length. Samples were conditioned at room temperature for at least 48 h before testing and tested using a programmed elongation rate of 50 mm min^{-1}. Load, displacement and strain were recorded during the test by means of an acquisition system connected to a computer. True stress σ and true strain ε were given by the relationships [12]:

$$\varepsilon = \ln\left(\frac{L}{L_0}\right) \qquad (2)$$

$$\sigma = \frac{F}{S} \qquad (3)$$

Using the law of conservation of volume ($SL=S_0L_0$), the true stress is written:

$$\sigma = \frac{FL}{S_0 L_0} \qquad (4)$$

where F, L_0, S_0, L and S were the applied force, the initial sample length, the initial sample section, the length and the sample section during the test, respectively.

2.2.3. Dynamic Mechanical Analysis. The viscoelastic properties of the polymers were determined via Dynamic Mechanical Analysis (DMA) using an apparatus DMA RSA3 from TA Instruments working in the tensile mode. Rectangular strips with dimensions around 40 x 5 x $1mm^3$ from all nanocomposite films as well as the neat NR film were prepared. Measurements were carried out each two degrees from -100 °C to 250 °C at a frequency of 1 Hz. The properties determined under this oscillating loading conditions were storage modulus (E'), loss modulus (E'') and damping factor ($\tan\delta$). In DMA, E' is a measure of stored energy without phase difference between stress and strain and represents the elastic component of a viscoelastic material; E'' is a measure of the energy lost as heat and represents the viscous component of a viscoelastic material; δ represents the phase lag between the applied stress and the corresponding strain in a viscoelastic material, while $\tan\delta$ is the ratio of loss modulus to storage modulus (E''/E') and is a measure of the energy dissipation or damping. $\tan\delta$ reaches a maximum as the polymer undergoes a transition from the glassy to the rubbery state at the glass transition temperature, T_g.

3. Results and discussion

3.1. Dielectric properties

Fig.1.a. and Fig.1.b. show the temperature dependence of the dielectric permittivity ε' and loss factor ε'', respectively, for the NR matrix. Dielectric measurements present three temperature regions of response assigned to:

(*i*) the primary relaxation associated with the glass transition and characterizing the α dipolar dynamic relaxation [13].

(*ii*) the water dipoles relaxation. Since NR has no hydrophilic groups, this observation suggests that water is connected to the hydroxyl groups (−OH) of the lipid present in NR [9].

(*iii*) the ionic conduction process appearing for high temperature and low-frequency ranges, which arises from the increase in the mobility of the electric charges in the polymer with temperature, with a large increase in both the real and imaginary parts of the dielectric function [14, 15].

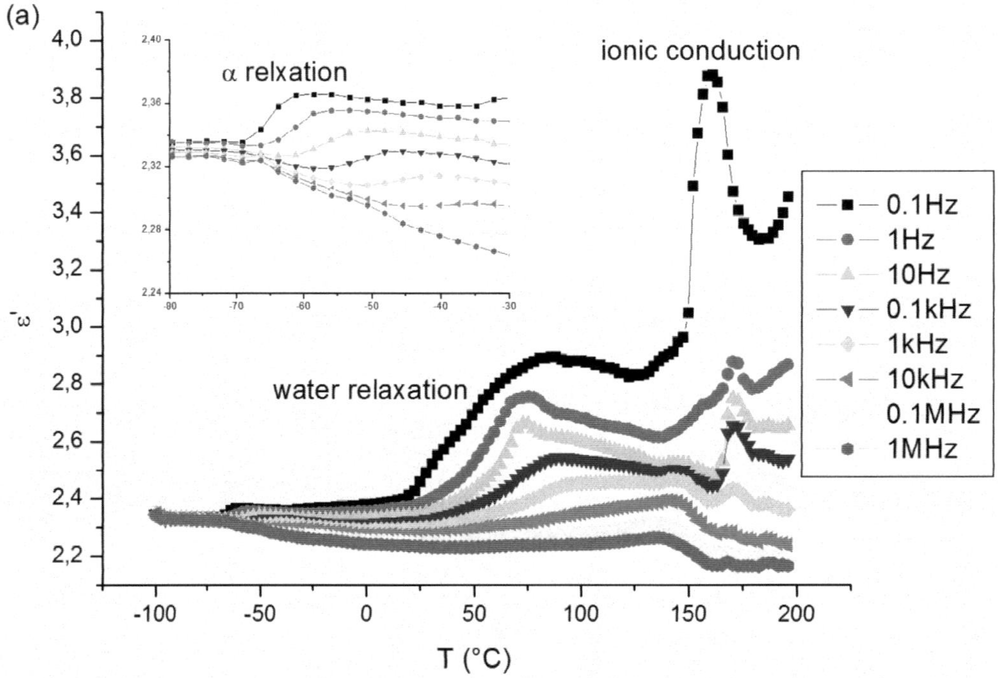

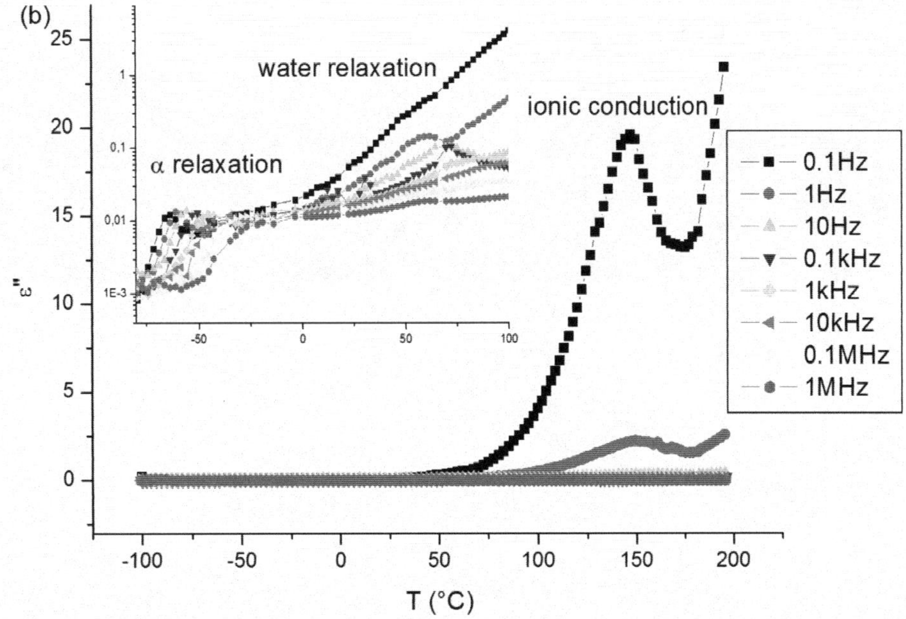

Fig. 1. Real and imaginary parts of the complex permittivity versus temperature, for the NR matrix.

In Fig. 2, the incorporation of cellulose nanofillers has an amplification effect of the dielectric losses. The polarity of cis-1,4-isoprene molecules is very low, so the loss data will be strongly influenced by dipolar impurities along the chains and in the bulk. The high value of the imaginary part of the permittivity of the nanocomposite material should be also due to the presence of water and impurities present on the cellulose nanofillers. The water dipoles relaxation is increasing when introducing hydrophilic cellulose. In addition, an interfacial polarization was detected. The interfacial polarization effect arises from the fact that free charges (catalysts, impurities, etc.), which were present at the stage of the processing, are immobilized in the material. These free carriers are then blocked at the interface between the cellulose nanofillers and NR matrix [16].

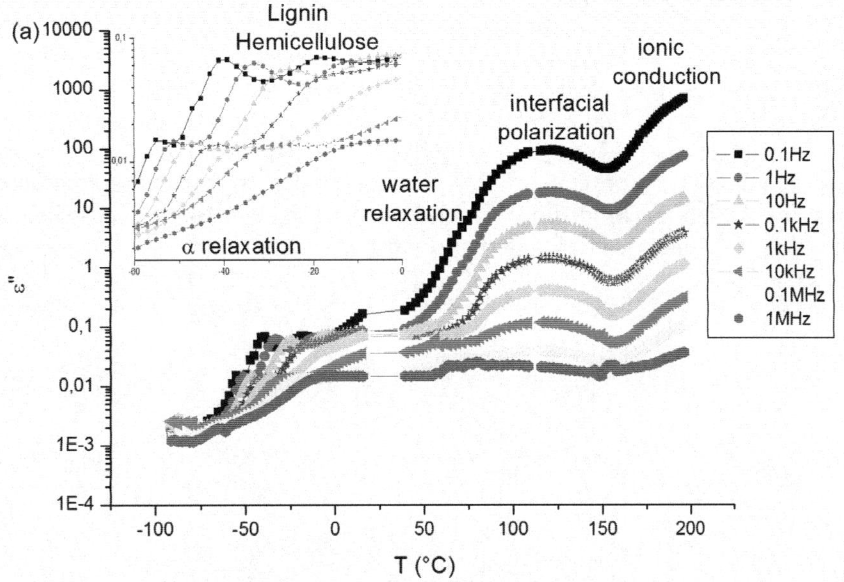

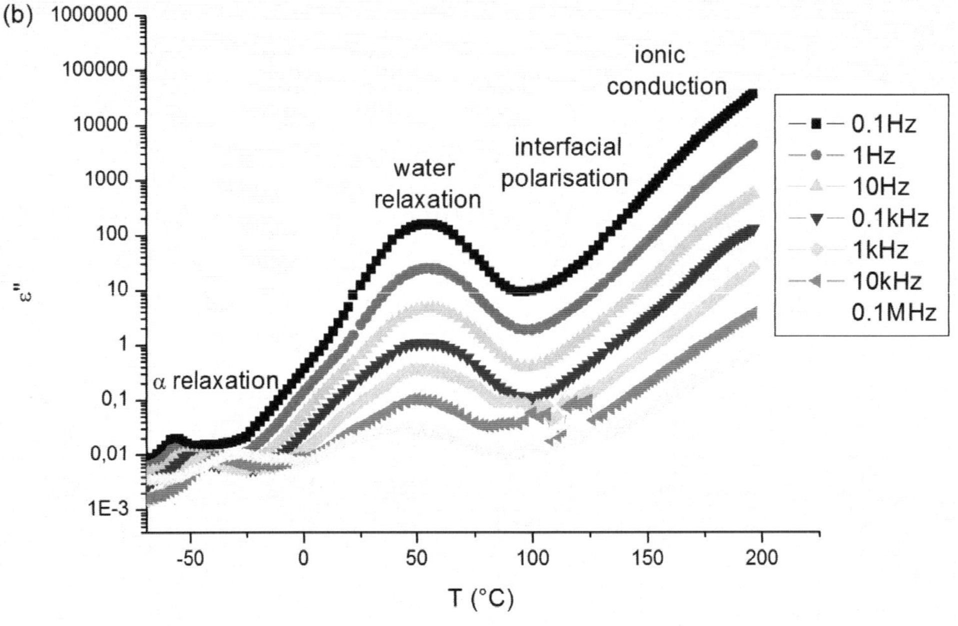

Fig. 2. Imaginary part of the complex permittivity, as a function of temperature, for (a) the NR-NFC7.5 and (b) NR-CNC7.5 nanocomposite.

In the dielectric study, isothermal runs give more information, and relaxation phenomena are investigated at the studied frequency range. However, relaxation peaks are often hidden with the large increase in permittivity values and loss at low-frequencies. The familiar difficulties of the ionic conduction phenomenon can be resolved using the first derivative of the real part of the dielectric permittivity [17]:

$$\varepsilon''_{der} = -\frac{\pi}{2}\frac{\partial \varepsilon'}{\partial \ln \omega} \qquad (5)$$

As the $\varepsilon'(\omega)$ values are in principle not affected by the Ohmic conductivity, and according to the Kramers-Kroning relationships, its derivative is proportional to the part of $\varepsilon''(\omega)$ resulting from dipolar processes [18]. Indeed, in the derivative formalism, the relaxation processes visible in the dielectric loss data appear as sharper peaks and without the conductivity contribution [19].

In Fig. 3, we compare the dielectric loss derivative data ε''_{der} at 150 °C for the NR, NR-NFC7.5 and NR-CNC7.5 films. For the NR matrix, we identify tow relaxation peaks attributed both to the relaxation of the lipid portion present in NR as a natural impurity [20, 21]. Lipid groups appeared on the NMR spectrum of NR performed in Ortiz-Serna work [20]. For nanocomposite samples, the lipid dipoles relaxations are hidden with the interfacial polarization.

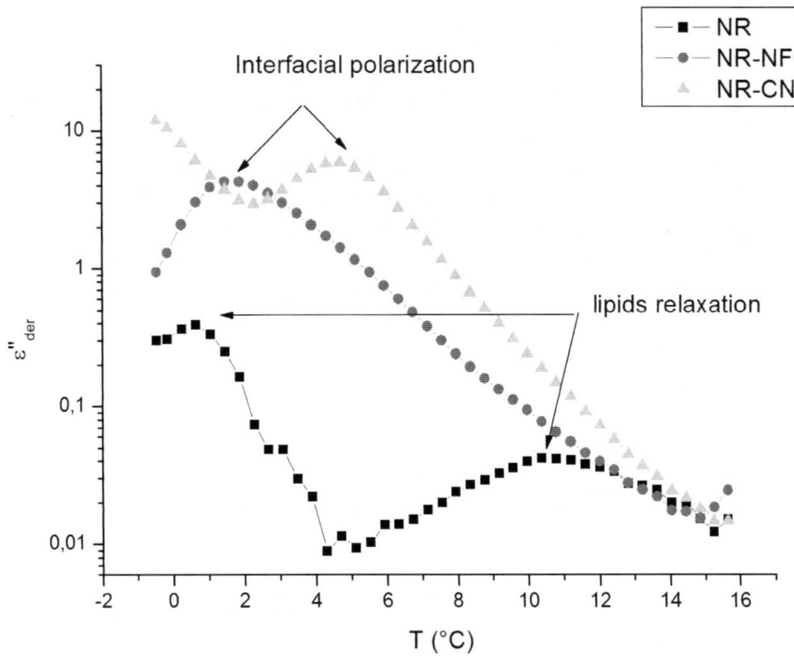

Fig. 3. The first derivative of the real part of the dielectric permittivity versus frequency at 150 °C.

To evaluate the effect of the reinforcement, the interfacial polarization (informing about the quality of the interface) is investigated. An analysis based on fitting using appropriate model functions is performed. The dielectric spectra (isothermal runs) were fitted by a sum of Havriliak-Negami (HN) model function terms of the form [22, 23]:

$$\varepsilon^*(\omega) = -i\left(\frac{\sigma_0}{\varepsilon_0\omega}\right)^N + \sum_{i=1}^{n}\left[\left(\varepsilon_{i\infty} + \frac{\Delta\varepsilon_i}{\left[1+(i\omega\tau_i)^{\alpha_i}\right]^{\beta_i}}\right)\right] \quad (6)$$

Where the term $-i\left(\frac{\sigma_0}{\varepsilon_0\omega}\right)^N$ is the contribution from polarization effects at the electrode, σ_0 is the DC conductivity, ε_0 is the permittivity of free space, and $\omega = 2\pi f$ is the radian frequency and N is a fractional exponent. Electrode polarization is ohmic when $N = 1$. The term $\sum_{i=1}^{n}\left[\left(\varepsilon_{i\infty} + \frac{\Delta\varepsilon_i}{\left[1+(i\omega\tau_i)^{\alpha_i}\right]^{\beta_i}}\right)\right]$ is the contribution of the dipolar orientations which are summed for the number of relaxation processes being fit simultaneously. $\varepsilon_{i\infty}$ is the real permittivity at $\omega = \infty$ of the *i*th relaxation process; $\Delta\varepsilon_i = \varepsilon_{si} - \varepsilon_{\infty i}$, is the dielectric relaxation strength where ε_{si} is the real permittivity at $\omega = 0$; τ_i is the relaxation time of the i_{th} relaxation process; α_i and β_i are the fractional shape parameters that respectively define the breadth and symmetry of the i_{th} relaxation process.
The polarization strength $\Delta\varepsilon$ is defined for static electric fields as [24–27]:

$$\Delta\varepsilon = \frac{4\pi F N_A \rho g \mu^2}{3\varepsilon_0 M_w k_B T} \quad (7)$$

Here $F\left(=\frac{\varepsilon_s(\varepsilon_\infty+2)^2}{3(2\varepsilon_s+\varepsilon_\infty}\right)$ is the Onsager factor for local field effects, N_A is Avogadro's number, M_w is the molecular weight, ρ is the density, k_B is Boltzmann's constant, g is the Kirkwood–Frohlich correlation factor and μ^2 is the mean squared dipole moment in the gas phase.

According to the table 2, the high increase in the polarization strength $\Delta\varepsilon$ with temperature is explained with the increase in the induced dipoles mobility. In addition, at a fixed temperature, $\Delta\varepsilon$ increases with filling up to a threshold value and then decreases. In fact, total interface area is increased with filling. The number of trapped dipoles at the interface is increased, increasing thus $\Delta\varepsilon$. Above a critical filler contain, the interface becomes sufficiently charged, and the interactions between the dipoles become more important. $\Delta\varepsilon$ then decreases indicating a rigid interface. In the other hand, and as it can be seen from table 2, one can predict that the nanocomposite reinforced with CNC has higher values of $\Delta\varepsilon$ than which reinforced with NFC. In fact, the number of trapped charges at the interface NR/CNC is more important than NR/NFC. Strong interactions between induced dipoles located at the interface NR/CNC indicate stiff interfaces. Therefore, this result indicates that the interfacial adhesion should be higher for NR/NFC than NR/CNC. It could be ascribed to the presence of residual lignin and hemicelluloses at the surface of NFC [18] acting as a compatibilizer between the filler and the matrix.

Table 2. The interfacial polarization strength $\Delta\varepsilon$ for different nanocomposites. The correlation coefficient, in all the cases, is larger than 0.997.

T (°C)	NR-NFC1	NR-NFC2.5	NR-NFC7.5	NR-NFC15
110	2.99	4.71	14.37	0.12
150	2.34	50.56	65.82	39.55
190	-	56.19	112.75	97.70
	NR-CNC1	NR-CNC2.5	NR-CNC7.5	NR-CNC15
110	12.33	40.65	30.12	12.84
150	36.52	85.13	79.33	54.32
190	80.64	190.25	155.95	101.60

3.2. Tensile properties

Tensile tests were run in order to evaluate the interactions between the fillers and the polymer matrix. Extracted parameters derived from Stress-Strain tensile tests are reported in Table 3.

Table 3. Young modulus (E), strain at break (ε_R) and strength (σ_R).obtained from the tensile tests [8].

Samples	E (MPa)	σ_R (MPa)	ε_R (%)
NR	0.5	0.56	575
NR-NFC1	1.27	0.7	209
NR-NFC2.5	10.52	0.8	14.6
NR-NFC7.5	121.2	4.15	8.18
NR-NFC15	233	6.26	3.95
NR-CNC1	1.7	0.86	408
NR-CNC2.5	2.8	1.17	358
NR-CNC7.5	59.2	5.7	170
NR-CNC15	187	12.15	14

For both sets of nanocomposites, the elongation at break decreases and both the tensile modulus and the strength increase drastically when increasing the nanoparticles content. The tensile modulus is systematically higher for a given composition for NFC reinforced nanocomposites compared to CNCs, whereas both the strength and the elongation at break are lower. These results seem in discordance with those of Ismail et al. [28] and Zhong Xian Ooi et al. [29] whose studied the reinforcement of natural rubber by carbon particles. The authors reported that a smaller particle size normally gives better interaction with the natural rubber matrix. The carbon black has the smallest particle size, thus the carbon black–filled NR vulcanizates showed the highest tensile strength and the lower strain at break followed by oil palm ash (OPA) and silica–filled vulcanizates. The presence of residual lignin and hemicelluloses at the surface of NFC could probably explain this discordance. Entangled NFC provide a higher stiffness than CNCs but induce brittleness of the material. Similar results were reported for chemically modified cellulose CNCs and MFC extracted from sisal reinforced polycaprolactone [30]. The presence of these impurities explains the higher compatibility of NFC with the hydrophobic NR matrix, and the ensuing higher filler/matrix adhesion. This higher NFC reinforcing effect was already evidenced from the dielectric study.

3.3. Dynamic Mechanical properties

The thermo–mechanical behavior of NR nanocomposites was studied by (DMA) analysis. Compared to tensile tests, this technique allows following the mechanical properties in a large temperature range. Fig. 4(a) shows the evolutions of the storage tensile modulus, log (E'), and the loss angle tanδ versus temperature at 1 Hz for NR, NR-NF2.5 and NR-CNC2.5. Modulus values have been normalized at low temperature at 1 GPa to allow a better comparison. The curve corresponding to the unfilled NR matrix is typical for high molecular weight amorphous thermoplastic polymer. The sharp decreases over 3 decades observed around -60 °C corresponds to the main relaxation phenomenon. It occurs in the same temperature range as T_g and it is therefore associated with the inelastic manifestation of the glass transition. The global evolution for nanocomposite films is similar to the one of the neat matrix. At low temperature, i.e., below T_g, the reinforcing effect of cellulosic nanofillers was low justifying the normalization of the modulus. Above T_g, a much more significant reinforcing effect of the nanofillers was observed. The NR-NFC2.5 nanocomposite has a higher E' than the NR-CNC2.5 nanocomposite, suggesting that the interactions between the NR matrix and the NFC fibers were stronger. As explained in the tensile test part, this could be ascribed to the presence of residual lignin and hemicelluloses at the surface of NFC [31] acting as a compatibilizer between the nanofiller and the matrix.

This result is confirmed in tanδ curves (Fig 4). As the dielectric study, only for the NR-NFC2.5 nanocomposite, the main relaxation process splits into two well–defined peaks. This splitting of the relaxation process could be ascribed to strong interactions between NFC and NR matrix because of the presence of residual lignin and hemicelluloses at the surface of NFC as already suggested. These interactions could lead to the formation of an interfacial layer surrounding the filler and which mobility is restricted compared to the bulk matrix. In addition, the NR matrix has the highest tanδ value indicating that there is a large degree of mobility, thus good damping behavior [46]. Reinforcement of NR with NFC or CNC nanofillers decreased the damping behavior of the nanocomposites as cellulose acted as a barrier to the free movement of the molecular chain. In addition, cellulose nanocomposites show lowest tanδ peak value and lowering of peak value also indicative of good interfacial adhesion [32].

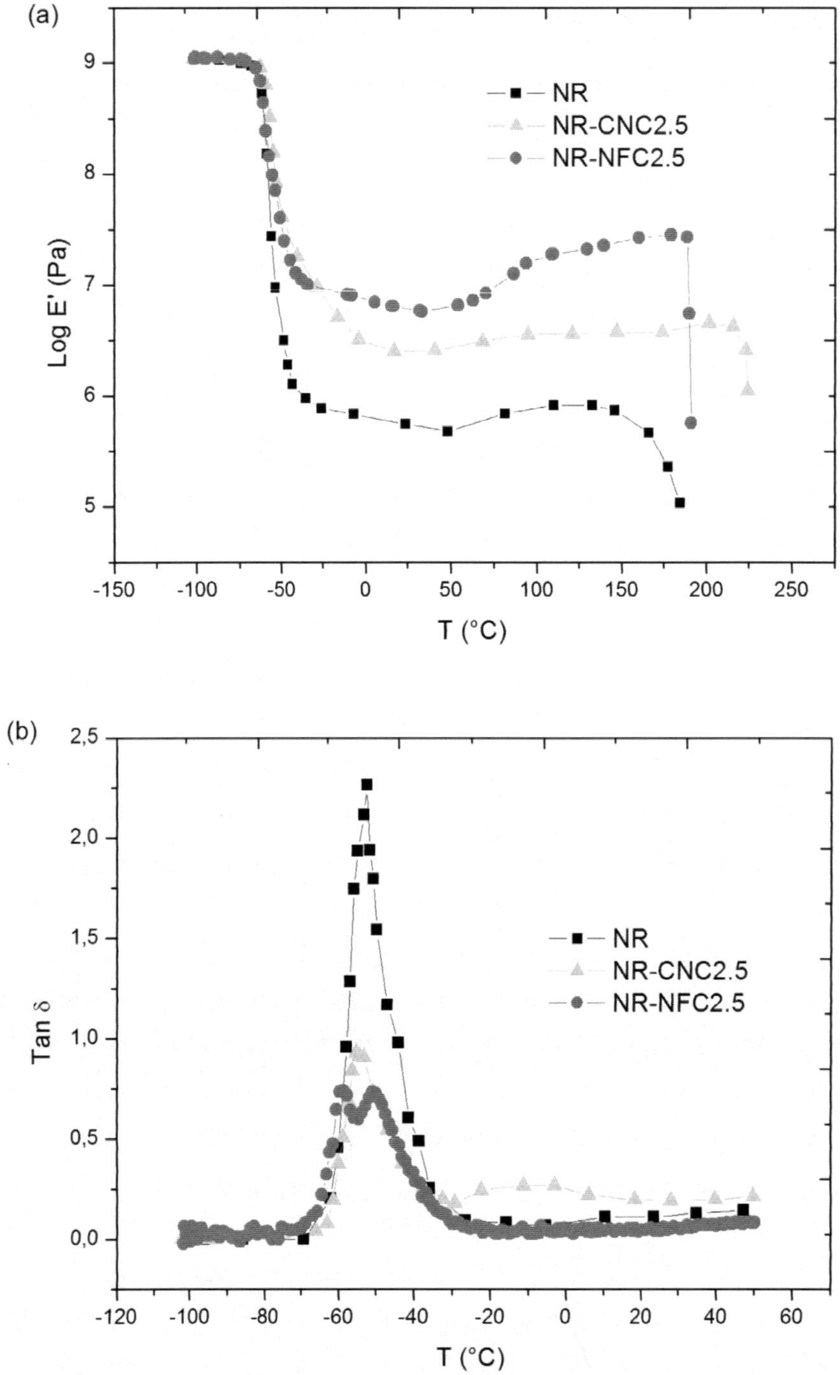

Fig. 4. Logarithm of the storage modulus (E') and tanδ curves vs. temperature for NR matrix, NR-NFC2.5 and NR-CNC2.5 nanocomposites.

These results clearly highlight that the interfacial adhesion is higher for NR/NFC nanocomposite compared to NR/CNC. This difference is manifested by the presence of a second relaxation attributed to constrained interfacial NR chains interacting with lignin (Scheme 1).

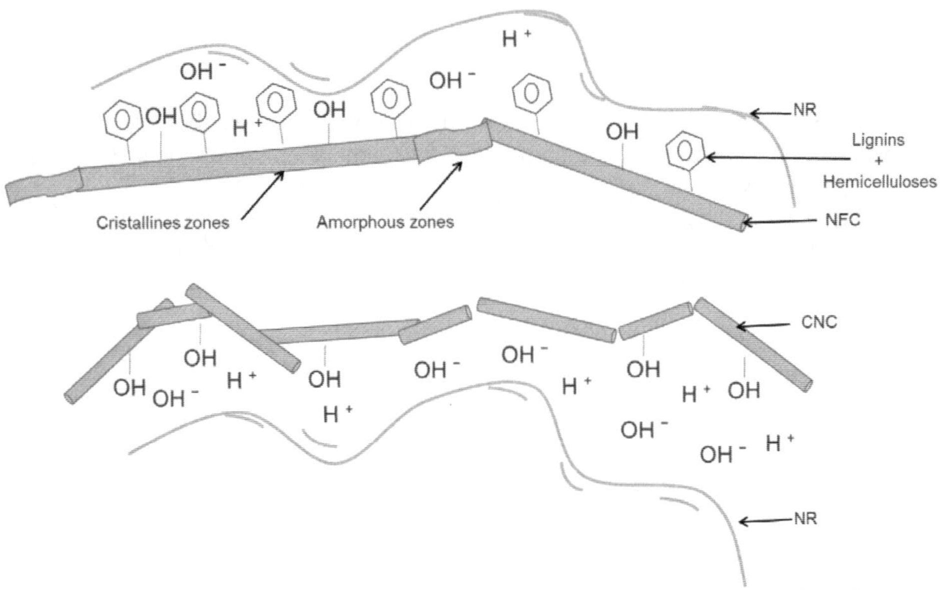

Scheme 1. Schematic illustration of the NR–NFC and NR–CNC interfaces.

4. Conclusion

This work emphasizes the cellulose nanofillers effect on the physical properties of NR based bio-nanocomposites. It attempts to give an overview of ongoing studies on dielectric, mechanical and dynamic mechanical properties of NR/NFC and NR/CNC nanocomposites. Being able to control the interface makes it possible to achieve the greatest improvement in the dielectric properties without strongly affecting the mechanical properties. By studying two different types of fibers, it was found that residual lignin and hemicelluloses surrounding NFC increase the tensile modulus and decrease the deformation at break. These impurities thus improve the interfacial adhesion between fibers and matrix. Moreover, the storage modulus E' is higher and the damping factor tanδ is lower for NR-NFC2 compared to NR-CNC2.5. This result suggests that the interactions between the NR matrix and NFC fibers were stronger. In addition, fiber size influences the dielectric properties of the resulting nanocomposites. Cellulose nanocrystals are smaller than nanofibrillated cellulose and CNCs generate more interfaces. Strong interactions between induced dipoles located at the NR/CNC interface indicate stiff interfaces. Therefore, the interfacial adhesion should be higher for NFC/NR than for NR/CNC.

References

[1] Boufi S and Chaker A 2016 Easy production of cellulose nanofibrils from corn stalk by a conventional high speed blender *Ind. Crop. Prod.* **93** 39-47.
[2] Ladhar A, Arous M, Kaddami H, Raihane M, Kallel A, Graça M P F and Costa L C 2014 Molecular dynamics of nanocomposites natural rubber/cellulose nanowhiskers investigated by impedance spectroscopy *J. Mol. Liq.* **196** 187-191.
[3] Benhamou K, Kaddami H, Magnin A, Dufresne A and Ahmad A 2015 Bio-based polyurethane reinforced with cellulose nanofibers: A comprehensive investigation on the effect of interface *Carbohyd. Polym.* **122** 202–211.
[4] Kamisan A S, Kudin T I T, Ali A M M and Yahya M Z A 2009 Gel polymer electrolyte based on methyl-grafted natural rubber for proton batteries *Mater. Res. Innov.* **13** 263–265.
[5] Chiappone , Nair J R, Gerbaldi C, Jabbour L, Bongiovanni R, Zeno E, Beneventi D and Penazzi N 2011 Microfibrillated cellulose as reinforcement for Li-ion battery polymer electrolytes with excellent mechanical stability, *J. Power Sources* **196** 10280–10288.
[6] Ummartyotin S, Manuspiya H, 2015 A critical review on cellulose: from fundamental to an

approach on sensor technology, *Renew. Sustain. Energy Rev.* **41** 402–412.

[7] Xu Q, Kong Q, Liu Z, Wang X, Liu R, Zhang J, Yue L, Duan Y and Cui G 2014 Cellulose/polysulfonamide composite membrane as a high performance lithium-ion battery separator *ACS Sustain. Chem. Eng.* **2** 194–199.

[8] Bendahou A, Kaddami H and Dufresne A (2010) Investigation on the effect of cellulosic nanoparticles' morphology on the properties of natural rubber based nanocomposites *Eur. Polym. J.* **46** 609–620.

[9] Dufresne A 2006 Comparing the mechanical properties of high performances polymer nanocomposites from renewable sources *J. Nanosci. Nanotechnol.* **6** 322–330.

[10] Dufresne A 2008 Polysaccharide nanocrystals reinforced nanocomposites *Can. J. Chem.* **86** 484–494.

[11] Arous M, Ben Amor I, Kallel A, Fakhfakh Z and Perrier G 2007 Crystallinity and dielectric relaxations in semi-crystalline poly(ether ether ketone) *J. Phys. Chem. Solids* **68** 1405–1414.

[12] Mariano M, El Kissi N and Dufresne A 2016 Cellulose nanocrystal reinforced oxidized natural rubber nanocomposites *Carbohydr. Polym.* **137** 174–183.

[13] Ladhar A, Arous M, Kaddami H, Raihane M, Lahcini M, Kallel A, Graça M P F and Costa L C 2013 Dielectric relaxation studies on nanocomposites of rubber with nanofibrillated cellulose *J. Non-Cryst. Solids* **378** 39–44.

[14] Ladhar A, Arous M, Kaddami H, Raihane M, Kallel A, Graça M P F, Costa L C 2015 Ionic hopping conductivity in potential batteries separator based on natural rubber–nanocellulose green nanocomposites *J. Mol. Liq.* **211** 792–802.

[15] Ladhar A, Arous M, Kaddami H, Raihane M, Kallel A, Graça M P F, Costa L C 2015 AC and DC electrical conductivity in natural rubber/nanofibrillated cellulose nanocomposites *J. Mol. Liq.* **209** 272–279

[16] Tsonos C, Apekis L, Zois C, Tsonos G 2004 Microphase separation in ion– containing polyurethanes studied by dielectric measurements *Acta Mater.* **52** 1319–1326.

[17] Kremer F, Schönhals A 2003 Broadband Dielectric Spectroscopy, Springer-Verlag, Berlin.

[18] Wübbenhorst M and Van Turnhout J 2002 Analysis of complex dielectric spectra. I. One-dimensional derivative techniques and three-dimensional modelling *J. Non-Cryst. Solids* **305** 40–49.

[19] Wübbenhorst M, Van Koten E, Jansen J, Mijs W and Van Turnhout J 1997 Dielectric relaxation spectroscopy of amorphous and liquid crystalline side chain polycarbonates *Macromol. Rapid Commun.* **18** 139.

[20] Ortiz-Serna P, Dıiaz-Calleja R, Sanchis M J, Floudas G, Nunes R C, Martins A F and L L Visconte 2010 Dynamics of Natural Rubber as a Function of Frequency, Temperature, and Pressure. A Dielectric Spectroscopy Investigation *Macromolecules* **43** 5094–5102.

[21] Ortiz-Serna P, Carsí M, Redondo-Foj B and Sanchis M J 2014 Electrical conductivity of natural rubber–cellulose II nanocomposites *J. Non-Cryst. Solids* **405** 180–187.

[22] Havriliak S, Negami S 1966 A complex plane analysis of α-dispersions in some polymer systems *J. Polym. Sci. C* **14** 99–117.

[23] Havriliak S and Negami S 1967 A complex plane representation of dielectric and mechanical relaxation processes in some polymers *Polymer* **8** 161–210.

[24] Langevin P 1905 Magnétisme et théorie des électrons *Ann. Chim. Phys.* **5** 70–127.

[25] Debye P 1929 Polar Molecules, The Chemical Catalogue Company, New York.

[26] Onsager L 1936 Electric Moments of Molecules in Liquids *J. Am. Chem. Soc.* **58** 1486–1493.

[27] Frohlich H 1958 Theory of Dielectrics, Oxford University Press, Oxford.

[28] Ismail H, Rozman, H D, Jaffri R M and MohdIshak Z A 1997 Oil palm wood flour reinforced epoxidized natural rubber composites: the effect of filler content and size *Eur. Polym. J.* **33** 1627–1632.

[29] Ooi Z X, Hanafi I and Abu Bakar A 2013 Optimisation of oil palm ash as reinforcement in natural rubber vulcanisation: A comparison between silica and carbon black fillers. *Polymer*

IOP Conf. Series: Materials Science and Engineering **258** (2017) 012001 doi:10.1088/1757-899X/258/1/012001

Testing **32** 625–630.
[30] Marchessault R H, Morehead F F and Walter NM 1959 Liquid crystal systems from fibrillar polysaccharides *Nature* **184** 632–633.
[31] Dufresne A, Cavaillé J Y and Vignon M R 1997 Mechanical behavior of sheets prepared from sugar beet cellulose microfibrils *J. Appl.Polym. Sci.* **64** 1185–1194.
[32] Jawaid M., Abdul Khalil H P S and Alattas O S 2012 Woven hybrid biocomposites: Dynamic mechanical and thermal properties *Composites: Part A* **43** 288–293.

IOP Conf. Series: Materials Science and Engineering **258** (2017) 012002 doi:10.1088/1757-899X/258/1/012002

Synthesis and characterization of GO-hydrogels composites

J Y Pereyra[1a], E A Cuello[1], R Coneo Rodriguez[1], C A Barbero[1], E I Yslas[3], H J Salavagione[2], D F Acevedo[1]*

[1] Departamento de Química, Facultad de Ciencias Exactas, Fisicoquímicas y Naturales, Universidad Nacional de Río Cuarto, 5800, Río Cuarto, Argentina.
[2] Dept. Polymer Physics, Elastomers & Energy Applications. CSIC-ICTP, C/ Juan de la Cierva, 3. 28006. Madrid, Spain.
[3] Departamento de Biología Molecular, Facultad de Ciencias Exactas, Fisicoquímicas y Naturales, Universidad Nacional de Río Cuarto, 5800, Río Cuarto, Argentina.

* dacevedo@exa.unrc.edu.

Abstract. The preparation of poly(N-isopropylacrylamide) (PNIPAm) hydrogel nanocomposites containing graphene oxide (GO) and GO plus carbon nanotubes (CNT) in the polymer network is communicated. This one-pot preparation methods include the dispersion of GO (or GO plus CNT) in a solution of monomers and the subsequent polymerization. The texture of the nanocomposites was studied using scanning electron microscopy (SEM), where very compact surfaces are observed suggesting good dispersion of GO sheets and CNTs within the polymer matrix. The presence of GO inside the polymer network diminished the equilibrium swelling values and increased the elastic modulus up to 162 % with respect to the pure gel. Similar results were observed for the composite with CNT. Furthermore, the electrical resistivity of PNIPAm-GO diminishes as the applied compression force increases, being 50 % lower than hydrogel without GO. Moreover, the electrochemical properties of the hydrogels, evaluated by cyclic voltammetry, indicate highly reversible electrical charge/discharge response. In order to apply these materials for antibiotic delivery, the absorption of tetracycline (tet) is evaluated and the nanocomposites showed better absorption capability and improved antibiotic delivery. Preliminary results suggest that tet loaded PNIPAm-GO and PNIPAM-GO-CNT display antimicrobial activity against the *Pseudomonas aeruginosa* turning these materials as potential candidates for biomedical applications.

1. Introduction

Carbonaceous nanomaterials such as graphene, graphene oxide and carbon nanotubes, have been attracting interest due to their properties that make them useful for technological applications, such as catalysis, coatings, and bio-applications [1–5]. On the other hand, responsive hydrogels have been in the focus of scientific attention in the last few decades owing to their numerous properties like high water content and relatively good deformability [6,7]. Among the different kinds of hydrogels, Poly(N-isopropylacrylamide) (PNIPAm) is one of the most popular termo-responsive hydrogels due to its peculiar volume phase transition temperature at c.a. 34 °C [8,9]. Notwithstanding the above mentioned properties, these hydrogels present some limitations, mainly related to its poor mechanical strength.

In this work, we describe the incorporation of GO and GO-CNT with the goal to improve the properties of the hydrogel. It is observed that the composites with GO and GO-CNT enhance the hydrogel mechanical and electric properties and also increase the capacity to incorporate tetracycline antibiotic molecules (tet) that allow to inhibit bacterial growth. Even though the swelling of these composite materials is smaller than the pure hydrogel, the elastic modulus increases up to 160 % with the incorporation of GO. Also, the hydrogel conductivity is enhanced with the introduction of GO and GO-CNT allowing potential technological applications in the resistive sensors areas. Likewise, the composites cyclic voltammetry studies show that these materials could be used as electrochemical electrodes. Furthermore, it is demonstrated that composites allow loading up to 50 % more of tetracycline solution than the pristine hydrogel. Accordingly, PNIPAm-GO and PNIPAM-GO-CNT with tetracycline inhibit the bacterial growth as is demonstrated in the inhibition halo experiments.

2. Materials and Methods

2.1. Graphene oxide synthesis

Natural graphite powder (Aldrich) was oxidized using a modification of the Hummers method [10,11]. Briefly, graphite powder was added to concentrated H_2SO_4 (200 mL) immersed in an ice bath. Then, $KMnO_4$ were added gradually under constant stirring and cooling to keep the suspension temperature below 20 °C. The reaction was kept under stirring during 2 hours. Subsequently, 360 mL of distilled water were added ensuring that the temperature does not exceed 50 °C. Then 40 ml of H_2O_2 (30%) was added until the color of the mixture turned bright yellow. The mixture was washed with bidistilled water until the pH reached a value of 3. The supernatant was discarded and the graphite oxide obtained was dispersed in distilled H_2O and subjected to ultrasound to generate GO sheets during one hour.

2.2. Hydrogel and composites synthesis:

Hydrogels matrix used N-isopropylacrylacrylamide (NIPAm, 0.5 M) as monomer precursor and N,N-methylene bisacrylamide (BIS, molar ratio 0.02:1 monomer based) as crosslinking agent [12]. The free radical polymerization was initiated by a redox initiator system: ammonium persulfate (APS, 1 mg mL^{-1}) and tetramethylenediamine (TEMED, 10 μL mL^{-1}). The Poly-N-isopropylacrylamede (PNIPAm) was synthesized dissolving NIPAm and the crosslinked agent BIS in 3 mL of bidistilled water, the reaction was initiated previous degasification incorporating APS and TEMED. The reaction was carried out at 22 °C during 1 hour. Hydrogels were polymerized in cylindrical molds, cut and washed several times with distilled water and dried for 48 hours at room temperature and moderate vacuum (0.1 atm) for later use. The hydrogels loaded with nanomaterials were synthetized using the same chemical reagents adding GO (5 mg mL-1) to obtain PNIPAm-GO, and adding the same quantity of GO and CNT (0.34 mg mL^{-1}) into the synthesis media to obtain PNIPAm-GO-CNT. The composites were polymerized in cylindrical molds, cut and washed several times with distilled water following the procedure descript previously for PNIPAm. The composites where observed by SEM (FE-SEM HITACHI S4800) microscopy.

2.3. Hydrogel and composites characterization:

The hydrogels ability of absorb/desorb water it is known as swelling capacity [13]. To analyze the swelling kinetics, PNIPAm, PNIPAm-GO and PNIPAm-GO-CNT dry gels were weighed and then immersed in bidistilled water at room temperature. Then, the materials were removed, dried, weighed and placed back in the solution at regular time intervals. The mass was measured after each interval time. The procedure was repeated until the materials achieved a constant weight. The swelling ratio was calculated according to Equation 1, as a function of time.

$$\%S_w = \frac{w(t) - wo}{wo} \cdot 100 \qquad (1)$$

Where: w(t) represents the weight of hydrogel in swollen state at time *t* and wo is the weight of the dry hydrogel. All the measurements were performance by triplicate.

2.4. Mechanical assays: Elastic-viscoelastic behaviour by uniaxial compression assay

Uniaxial compression measurements were performed on individual discoidal hydrogels and composites in swollen state at room temperature. For these measurements, a homemade device was built to measure the Young module, under compression. The stress (σ, KPa) versus strain (ε) curve was built following the procedure described by Martinez et al. [7], and from the slope the Young module was calculated [14].

2.5. Measurements of electric resistance

All the measurements were conducted at 25 ± 3°C. A cylindrical gel of 10 mm in diameter and 15 mm in height was placed on an analytic digital balance (OHAUS Pioneer). On the extremes of the cylinder two electrodes were placed, and then different loads were imposed vertically to the gel. The electrodes were connected to a Siemens multimeter and the electrical resistance was measured for each compression state [15].

2.6. Electrochemical measurements

Cyclic voltammetry experiments were carried out in a conventional three electrode cell, controlled by a computerized potentiostat (Autolab). All potentials were quoted against the saturated Ag/AgCl electrode and platinum wire as contraelectrode, in all the measures the materials were cut in cylindrical pieces (10mm diameter and 2 mm of height) and were contacted using glassy carbon and at scan rate 50 mV s^{-1}. The support electrolyte used was H_2SO_4 (1 M).

2.7. Partition coefficient measurements by UV-vis spectroscopy

The CP value determines the distribution of tet between two phases, hydrogel and water and it is defined by the Equation 2.

$$CP = \frac{C_{material}}{C_{solution}} \qquad (2)$$

Where $C_{material}$ and $C_{solution}$ is the tetracycline molal concentration in the gels and tet molal concentration in the solution, respectively.

Experimentally, a piece of known mass dry materials was immersed in 3 mL of tet solution. After 8 days of immersion, the concentration of the remaining tet in the solution was determined by UV-visible spectroscopy (Hewlett-Packard- 8453 UV-visible Spectrophotometer). Previously, a calibration curve of tet solution was carried out fixing the wavelength at 360 nm. The volume of the hydrogel was evaluated by gravimetric measures, considering the solution density same to the water density and at 25 °C [16].

2.8. Antibacterial activity studies

The materials antimicrobial activity against *Pseudomonas aeruginosa* (ATCC) was tested according to agar diffusion test [17]. For agar diffusion method, the samples were exposed to bacteria in solid media (nutrient agar), and the inhibition zone around each sample was measured and recorded. This method was performed in Luria Bertani (LB) medium solid agar Petri dish. Luria Bertani medium was used as a growing medium of *Pseudomonas aeruginosa* bacteria in overnight inoculums. Then, 100μL of bacterial solution inoculum was placed over the agar into sterilized Petri dishes. The materials were placed on the *Pseudomonas aeruginosa* cultured agar plate. The agar plates were incubated for 24 h at 37 °C and the inhibition zone was monitored. After incubation, the presence of a bacterial growth inhibition halo around the samples was observed and their diameter in millimetres was measured. The assays were performed in triplicate.

3. Results and Discussion

Figure 1 shows digital images of PNIPAm hydrogel and its composites with GO and GO-CNT, where the black color, imparted by the fillers is clearly observed, confirming the presence of GO and CNT in the hydrogel network (Fig. 1 b and c.). SEM image of the pure PNIPAm hydrogel depicts a homogeneous polymer surface (Fig. 1d). However, in the case of PNIPAm-GO the surface presents a layered structure due to the GO sheets. The PNIPAm-GO-CNT displays a similar structure as PNIPAm-GO with additional filamentous structures owing to the presence of the CNT (Fig. 1f).

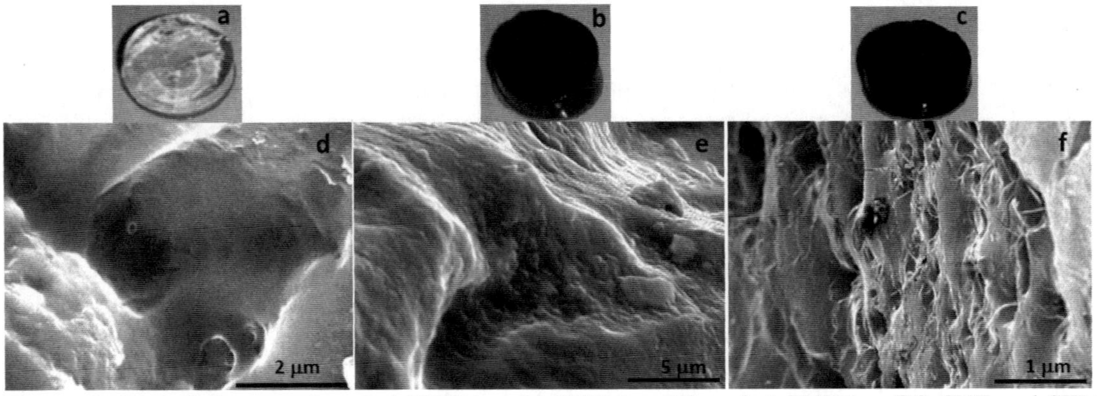

Figure 1: Digital photography of a) PNIPAm, b) PNIPAm-GO and c) PNIPAm-GO-CNT and SEM images of d) PNIPAm, e) PNIPAm-GO and f) PNIPAm-GO-CNT.

Figure 2 shows a comparison of the swelling kinetics between the hydrogel and the composites materials. The equilibrium swelling of PNIPAm is 3300 % and for the composites, the swelling percentage reaches lower values c.a. 1850 % for PNIPAm-GO and 2100 % for PNIPAm-GO-CNT. Considering that the composites equilibrium swellings are lower than that of the pure polymer it is reasonable to propose that their matrices are less elastic.

Regarding the mechanical performance, an increase the Young modulus has been observed when nanomaterials are incorporated: the PNIPAm-GO sample presented the highest value, 9.7 KPa, an enhancement of 160 % with respect to PNIPAm (3.7 KPa), resembling the case of other hydrogel nanocomposites [7]. This behavior could be explained considering two factors: *i)* the high rigidity of GO material (E = 290-430 GPa) [18], and *ii)* the GO homogeneous distribution within the hydrogel. Moreover, the groups -COOH, -OH and -C=O present in the GO surface are able to generate hydrogen bonding with the polymer groups, leading to greater rigidity. The composite with GO-CNT present a similar change in the equilibrium swelling that PNIPAm-GO, however the Young module value is slightly altered. These results could be explained considering that the GO-CNT in the matrix could produces a less pronounced decrease in the matrix elasticity due to the hydrophobic character of the CNT or a not a perfect homogeneous distribution of the CNT in the matrix.

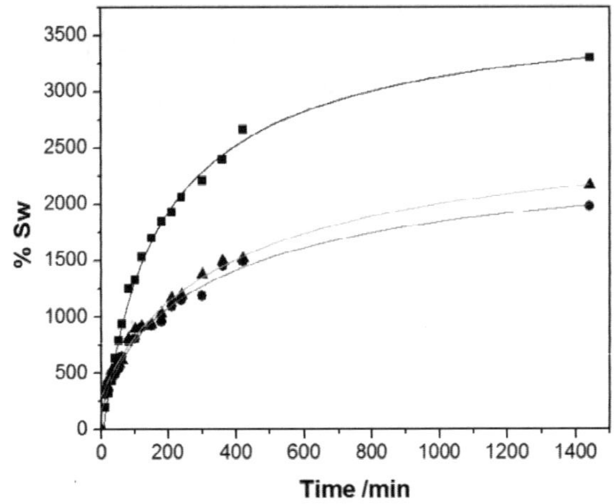

Figure 2: Swelling kinetic of (■) PNIPAm, (●) composite of PNIPAm with graphene oxide and (▲) PNIPAm with GO and CNT.

Table 1. Mechanical properties of the materials, Young modulus and ercentage of Young modulus increase

Materials	Young Modulus (KPa)	% of increase
PNIPAm	3.7 ± 0.1	0
PNIPAm-GO	9.7 ± 0.3	161.83
PNIPAm-GO-CNT	6.8 ± 0.3	82.96

The changes of the resistance at different loadings imposed vertically to the gel relative to the initial unloaded resistance are shown in Figure 3a. The PNIPAm hydrogels present higher values of resistance than the composites, as expected. In all cases the materials increase the conductivity as the applied load increase. The initial PNIPAm resistance at t = 0 (c.a 50 MΩ) is higher than PNIPAm-GO-CNT (45 MΩ) and NIPAm-GO (25 MΩ). It is possible to observe that the resistance diminishes when the vertical load applied increases, this is an expected behaviour, since the application of stress changes the spacing between conductive filler particles resulting in resistivity changes. The composites present lower resistivity values for all the load applicated, changin the value up to 0.6 for PNIPAm-GO. These results indicate that these composite materials are more sensitive to external pressure than the pure hydrogel. Beside the composite loaded with CNT present a higer change (0.7) this could be attibuited to the best conductivity of the CNT.

The cyclic voltammetry at a scan rate of 50 mV/s is shown in Figure 3b. The profile revealed a rectangular shape of the curves, which indicate highly reversible electrical charge/discharge response. All results indicate that the conductive behavior of these materials has been improved with the introduction of GO and with GO and CNT allow potentials technological application.

In order to apply these materials as inhibitors of bacterial growth, the partition coefficient (CP) of solutions of tet, an antibiotic used to treat several bacterial infections, was studied. Clearly, the composite materials present higher CP values (PNIPAm-GO 119; PNIPAm-GO-NTC 137) than PNIPAm (83), indicating that the GO and CNT improve the adsorption of the antibacterial agent (Figure 4a). This behaviour is in agreement with other studies that shown the GO ability to sorbe organic molecules [1,19]. To improve its use as antimicrobiane, the kinetics of liberation of tet from the hydrogels was evaluated (Figure 4b). As it can be seen in Figure 4b, the composites show enhanced delivery of the antibiotic in comparison with the pure hyrdogel.

IOP Conf. Series: Materials Science and Engineering **258** (2017) 012002 doi:10.1088/1757-899X/258/1/012002

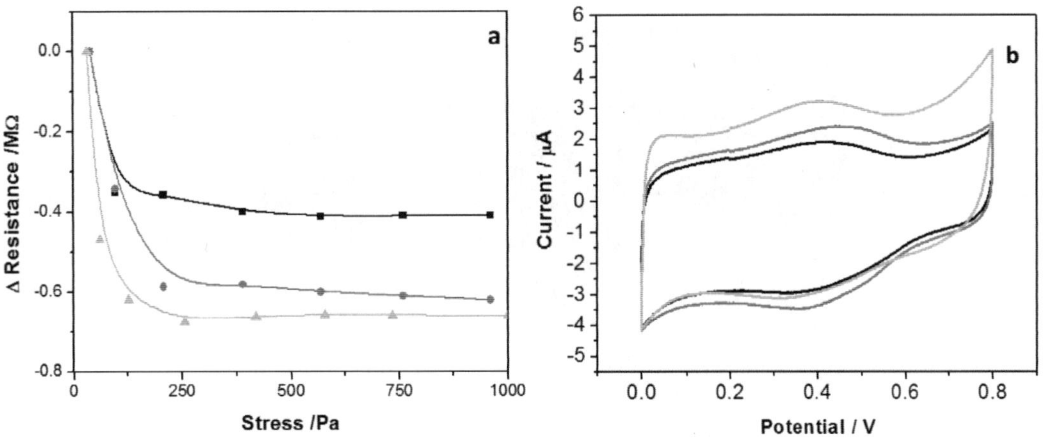

Figure 3: a) resistance variation of the materials versus stress (■) PNIPAM, (●) composite of PNIPAm with graphene oxide and (▲) PNIPAm with GO and CNT. b) Cyclic voltammetry of PNIPAm (black line), composite of PNIPAm with graphene oxide (dark gray) and PNIPAm with GO and CNT (light gray).

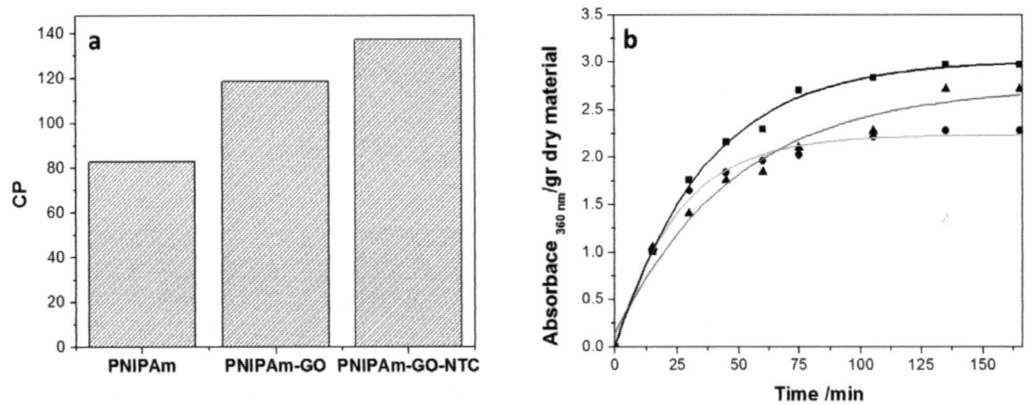

Figure 4: a) Partition coefficient of Tetraciclyne in PNIPAm hydrogels and PNIPAm-GO PNIPAm-GO-CNT composites and b) Kinetic of tet liberation of i) PNIPAm (light gray line, ●), ii) PNIPAm-GO (gray line ▲) and iii) PNIPAm-GO-CNT (black line ■).

The in vitro antibacterial properties of composite hydrogels were investigated by agar diffusion tests against gram-negative *Pseudomonas aeruginosa* bacteria. The inhibition zones are presented in Figure 5. The results clearly show that the inhibition zone around each loaded composite PNIPAm-GO-tet (1.60 ±0.2 cm) and PNIPAm-GO-CNT-tet (1.75±0.3 cm) are higher than the PNIPAm-GO and NIPAm-GO-CNT without the antibiotic which not present inhibition zone. The results suggest that the tet loaded composite hydrogels revealed a toxic effect on bacteria than hydrogel without the antibiotic under similar conditions as evidenced by higher inhibition zone due to the more efficiency delivery of the tet.

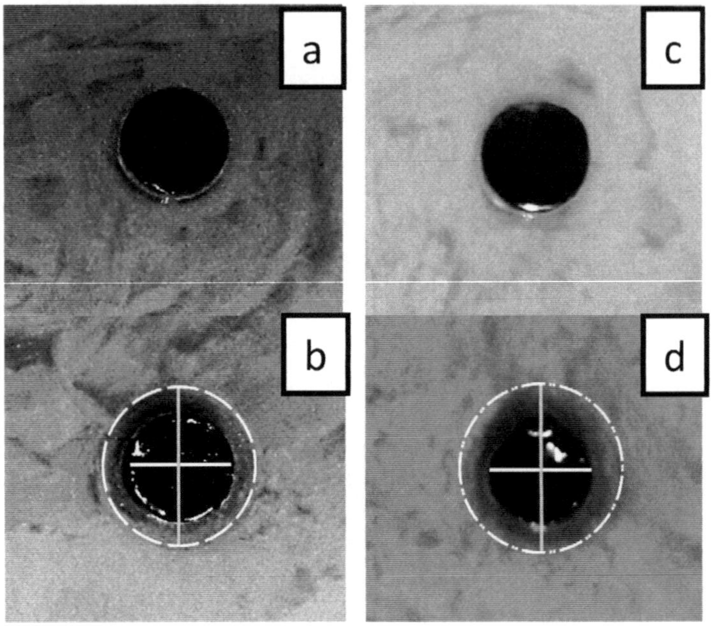

Figure 5: Photographs of the inhibition zones of nanocomposite hydrogels against *Pseudomonas aeruginosa*. a) composite PNIPAm with graphene oxide, b) composite PNIPAm with graphene oxide and tet, c) PNIPAm with GO and CNT and d) PNIPAm with GO and CNT and tet. The white line represents the hydrogel diameter and the yellow line shows the inhibition diameter.

4. Conclusion

Novel materials composed of a hydrogel (PNIPAm) matrix and GO or GO-CNT has been synthesized using a facile one-pot procedure. The nanomaterials incorporated in the matrix improve its mechanical properties. Even though the swelling degree of these composite materials is smaller than the pristine hydrogel, the elastic modulus is clearly enhanced. Furthermore, the hydrogel conductivity is enhanced with the introduction of GO and GO-CNT allowing potentials technological applications in resistive sensors. The electrochemistry behavior also opens a new application for these materials as e.g. selective electrodes. Moreover, the loading capacity to incorporate an antibiotic is around 50 % higher for the composites than for the pristine hydrogel. Besides, the agar diffusion test demonstrates the composite antimicrobial activity against the *Pseudomonas aeruginosa* of tet loaded composite hydrogels (PNIPAm with graphene oxide and tet and PNIPAm with GO and CNT and tet). These results indicate potential of tet loaded composite hydrogels for biomedical applications.

5. Acknowledgment

D.F. Acevedo E.I. Yslas, and C.A. Barbero, are permanent research fellows of CONICET. J. Y. Pereyra thanks FONCYT for a graduate fellowship. The funding of FONCYT (PICT 2013-2716), CONICET (PIP 2014-2016, No. 11220130100663CO), SECYT-UNRC and EU (IRSES project "SUMA2-Network", Pr. No: 318903) and MAT MAT2013-47898-C2-2-R (MINECO, Spain) is gratefully acknowledged. D Acevedo thanks to the German Academic Exchange Service (DAAD) and the European School of Materials for the financial support.

IOP Conf. Series: Materials Science and Engineering **258** (2017) 012002 doi:10.1088/1757-899X/258/1/012002

References

[1] Zhu Y, Murali S, Cai W, Li X, Suk J W, Potts J R and Ruoff R S 2010 Graphene and graphene oxide: Synthesis, properties, and applications *Adv. Mater.* **22** 3906–24

[2] Hu K, Kulkarni D D, Choi I and Tsukruk V V. 2014 Graphene-polymer nanocomposites for structural and functional applications *Prog. Polym. Sci.* **39** 1934–72

[3] Jeong H Y, Kim J Y, Kim J W, Hwang J O, Kim J E, Lee J Y, Yoon T H, Cho B J, Kim S O, Ruoff R S and Choi S Y 2010 Graphene oxide thin films for flexible nonvolatile memory applications *Nano Lett.* **10** 4381–6

[4] Levi-Polyachenko N, Young C, MacNeill C, Braden a, Argenta L and Reid S 2014 Eradicating group A streptococcus bacteria and biofilms using functionalised multi-wall carbon nanotubes *Int J Hyperth.* **30** 490–501

[5] Zhou F, Xing D, Ou Z, Wu B, Resasco D E and Chen W R 2009 Cancer photothermal therapy in the near-infrared region by using single-walled carbon nanotubes *J. Biomed. Opt.* **14** 21007–9

[6] Singh N K and Lee D S 2014 In situ gelling pH- and temperature-sensitive biodegradable block copolymer hydrogels for drug delivery *J. Control. Release* **193** 214–27

[7] Martínez M V., Bongiovanni Abel S, Rivero R, Miras M C, Rivarola C R and Barbero C A 2015 Polymeric nanocomposites made of a conductive polymer and a thermosensitive hydrogel: Strong effect of the preparation procedure on the properties *Polym. (United Kingdom)* **78** 94–103

[8] Kumar A, Srivastava A, Galaev I Y and Mattiasson B 2007 Smart polymers: Physical forms and bioengineering applications *Prog. Polym. Sci.* **32** 1205–37

[9] Sugiura S, Sumaru K, Ohi K, Hiroki K, Takagi T and Kanamori T 2007 Photoresponsive polymer gel microvalves controlled by local light irradiation *Sensors Actuators, A Phys.* **140** 176–84

[10] Hummers W S and Offeman R E 1958 Preparation of Graphitic Oxide *J. Am. Chem. Soc.* **80** 1339–1339

[11] Chen J, Yao B, Li C and Shi G 2013 An improved Hummers method for eco-friendly synthesis of graphene oxide *Carbon N. Y.* **64** 225–9

[12] Caykara T, Kiper S and Demirel G 2006 Thermosensitive poly(N-isopropylacrylamide-co-acrylamide) hydrogels: Synthesis, swelling and interaction with ionic surfactants *Eur. Polym. J.* **42** 348–55

[13] Baker J P, Stephens D R, Blanch H W and Prausnitz J M 1992 Swelling equilibria for acrylamide-based polyampholyte hydrogels *Macromolecules* **25** 1955–8

[14] Mulko L, Rivarola C R, Barbero C A and Acevedo D F 2016 Bioethanol production by reusable Saccharomyces cerevisiae immobilized in a macroporous monolithic hydrogel matrices *J. Biotechnol.* **233** 56–65

[15] Rivero R E, Molina M A, Rivarola C R and Barbero C A 2014 Pressure and microwave sensors/actuators based on smart hydrogel/conductive polymer nanocomposite *Sensors Actuators, B Chem.* **190** 270–8

[16] Molina M A, Rivarola C R and Barbero C A 2012 Study on partition and release of molecules in superabsorbent thermosensitive nanocomposites *Polymer (Guildf).* **53** 445–53

[17] Bonev B, Hooper J and Parisot J 2008 Principles of assessing bacterial susceptibility to antibiotics using the agar diffusion method *J. Antimicrob. Chemother.* **61** 1295–301

[18] Liu L, Zhang J, Zhao J and Liu F 2012 Mechanical properties of graphene oxides *Nanoscale* **4** 5910

[19] Potts J R, Dreyer D R, Bielawski C W and Ruoff R S 2011 Graphene-based polymer nanocomposites *Polymer (Guildf).* **52** 5–25

Thermal management of metallic surfaces: evaporation of sessile water droplets on polished and patterned stainless steel

T Czerwiec[1,2], S Tsareva[1,2], A Andrieux[1,2], G A Bortolini[1], P H Bolzan[1], G Castanet[3], M Gradeck[3] and G Marcos[1,2]

[1] Institut Jean Lamour, UMR CNRS 7198, Nancy, F-54042, Université de Lorraine, France.
[2] Laboratoire d'Excellence Design of Alloy Metals for low-mAss Structures (LabEx DAMAS), Université de Lorraine, Ile du Saulcy, F-57045 Metz cedex, France
[3]Laboratoire d'Energétique et de Mécanique Théorique et Appliquée, UMR 7563 NANCY, F-54042, Université de Lorraine, France.

* thierry.czerwiec@univ-lorraine.fr

Abstract. This communication focus on the evaporation of sessile water droplets on different states of austenitic stainless steel surfaces: mirror polished, mirror polished and aged and patterned by sputtering. The evolution of the contact angle and of the droplet diameter is presented as a function of time at room temperature. For all the surface states, a constant diameter regime (CCR) is observed. An important aging effect on the contact angle is measured on polished surfaces due to atmospheric contamination. The experimental observations are compared to a quasi-static evaporation model assuming spherical caps. The evolution of the droplet volume as a function of time is almost linear with the evaporation time for all the observed surfaces. This is in accordance with the model prediction for the CCR mode for small initial contact angles. In our experiments, the evaporation time is found to be linearly dependent on the initial contact angle. This dependence is not correctly described by the evaporation model.

1. Introduction

Heat production accounts for about 70% of the world's primary energy consumption. Better thermal management in industrial processes would be an advance in the energy transition process. In this context, the development of efficient extractors, heat exchangers, condensers, sufficiently compact and affordable constitutes an important industrial challenge. Studying the evaporation of droplets on a solid surface is a way to fulfil this objective.

When the temperature of a solid surface is lower than the saturation temperature of the liquid (T_{sat}), heat transfer is dominated by heat conduction from the substrate to the liquid, and evaporation occurs at the liquid-gas interface [1-3]. This evaporation regime is called "film evaporation". When the substrate temperature exceeds T_{sat}, the droplet boils and small bubbles nucleate at the liquid-solid interface. The corresponding evaporation regime is "nucleate boiling". However, a limiting heat flux, the so-called critical heat flux, is reached at a critical temperature, for which the total evaporation time of a droplet is at its minimum. When vapour bubbles burst at the liquid-gas interface, small droplets can be ejected from the drop. The corresponding heat transfer regime is the transition boiling regime [2]. For higher temperature, a thin micrometre-scale vapour layer interface formed between the water

IOP Conf. Series: Materials Science and Engineering **258** (2017) 012003 doi:10.1088/1757-899X/258/1/012003

droplet and the wall. The transition temperature between transition and film boiling is called the steady Leidenfrost temperature (T_{lste}). In many industrial processes, it is not desirable to work in the film boiling regime (above the Leidenfrost temperature) due to the inhibition of heat transfer. The Leidenfrost temperature thus limits the efficiency of metallurgical processes such as quenching and spray cooling, high heat flux microelectronics cooling systems and fire-fighting equipment [4].
An improvement of heat transfer performances can be obtained by enhancing the surface area. This can be achieved by roughening the surface, or by surface patterning. Surface patterning (surface structuration) allows the production of a "patterned" surface with some regular array of surface height features on the size scale of several micrometres to some nanometres. Robust and efficient surface patterning manufacturing methods are existing but alternative cheap and flexible technologies are required, especially for metallic surfaces. Plasma technologies such as nitriding, carburizing and etching by ion bombardment are well adapted technologies for that purpose [5-6]. Evaporation on textured surfaces is intensively studied, especially on textured surfaces having superhydrophobic properties [7-11]. However, it is not the case for plasma nitrided or plasma patterned by sputtering metallic surfaces. This communication will focus on the evaporation of sessile water droplets on different states of austenitic stainless steel AISI 316 L surfaces modified or not by plasma treatments: mirror polished and patterned by ion sputtering using masks. The experimental results will be compared to a simple quasi-static evaporation model assuming spherical caps for the evaporating droplets.

2. Experimental devices

Samples used in this communication are AISI 316L stainless steel cylinders (see table 1 for composition). The samples were cut from a cylindrical rod (20 mm of diameter) with a thickness of 5 mm. The samples were polished with a mirror-like finish that guarantee the parallelism between the two faces of each ring. The patterning treatment was done using a multi-dipolar electron cyclotron resonance system (MDECR) [5] in an argon-hydrogen gas mixture. The analysis of the traces on the samples was performed by optical microscopy. The confocal laser scanning microscope was a Zeiss LSM 700 (solid laser, 405 nm, 5 mW) and the spatial and time resolutions were respectively 70 μm (field of view 1248 × 1268 μm) and 0.64 fps. Contact angle measurements were performed with a Digidrop equipment at room temperature (22-23°C) with a humidity of 51-52%. Drops of ultrapure water were deposited with minimal influence of external forces and the process was recorded until the drop was fully evaporated, with 1 picture taken per second. The contact angle measurements were done without doing any surface cleaning before the test.

Table 1. Composition of the two AISI 316L stainless steel in wt%.

	Fe	Cr	Ni	Mn	Mo	Si	Cu	Co	N	P	C	S	N
AISI 316L sample 1	67	16.8	10.1	2.2	2	0.8	0.5	0.2	0.09	0.05	0.02	0.03	0.1
AISI 316LM sample 2	67.5	16.2	9.4	2.3	2	0.9	0.7	0.2	0.1	0.05	0.02	0.02	0.2

3. Evaporation of sessile droplet on stainless steel surfaces

3.1. The different evaporation regimes

Figures 1, 2 and 3 respectively show the evolution of the contact angle and drop width (2 R) as a function of time for polished austenitic stainless steel exposed to air for a long time, a patterned sample and polished austenitic stainless steel exposed to air after different ageing times. These figures illustrate the four possible evaporation regimes that can exist for a drop of liquid on a surface in the partial wetting situation [12-15]:

IOP Conf. Series: Materials Science and Engineering **258** (2017) 012003 doi:10.1088/1757-899X/258/1/012003

- Initial regime (propagation step): during this short initial spreading period (generally less than one minute), the contact angle and the radius R change simultaneously to reach the initial considered values for the next step (see the early beginning of figures 2 and 3).
- CCR Regime: the contact angle θ decreases while the radius R of the drop remains constant. This step is known as the constant contact radius evaporation regime (CCR). In this configuration, the triple line remains pinned (pinning mode).

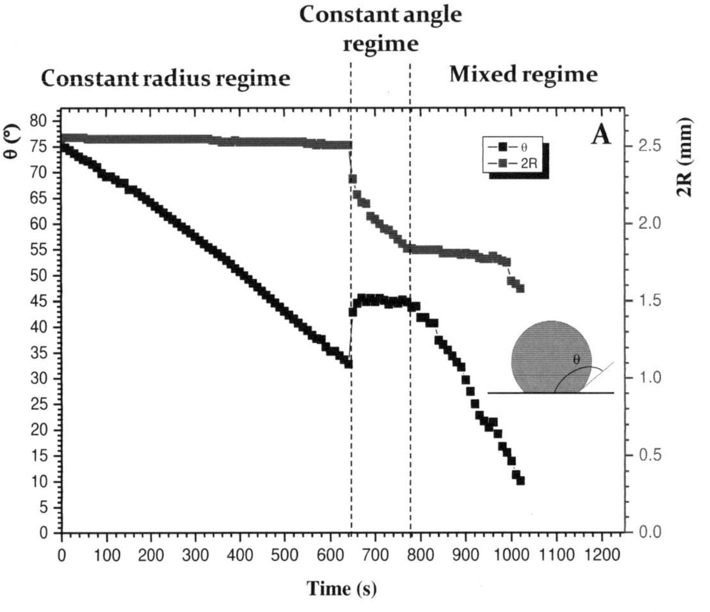

Figure 1. Evolution of the contact angle and drop width (2 R) as a function of time for a 3 μl water drop on polished stainless steel sample 1 exposed to air for a long time. The onset shows the definition of the contact angle.

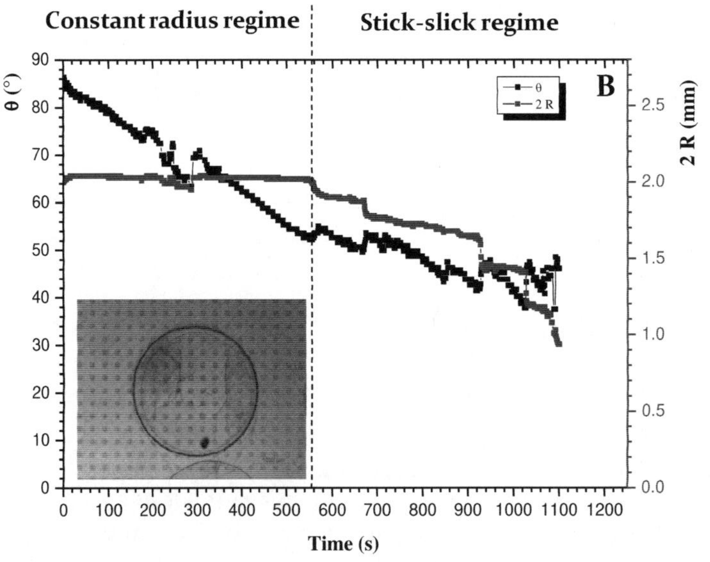

IOP Conf. Series: Materials Science and Engineering **258** (2017) 012003 doi:10.1088/1757-899X/258/1/012003

Figure 2. Evolution of the contact angle and drop width (2 R) as a function of time for a 1.9 µl water drop on a negatively patterned substrate sample 1 (squares of 100 × 100 µm separated by 200 µm from each other with a depth of 0.7 µm). The onset shows the trace remaining after the evaporation of a 3 mm in diameter water drop on the patterned surface.

- CCA Regime: the contact angle θ remains constant while the radius R of the drop decreases. This step is known as the constant contact angle evaporation regime (CCA). In this configuration, the triple line moves (depinning mode).
- Mixed Regime: the contact angle θ and the radius R of the drop decreases until the droplet completely disappears (mixed regime).
- Stick-slip regime: in this mode, consisting of one or more stages, the triple line is pinned (R is constant) and one or more "sliding" phases in which the triple line is moving (R varies) happens.

As seen on figure 1, three different regimes for the evaporation exist on austenitic stainless steel AISI 316L: a constant diameter regime, a constant contact angle regime and a mixt regime in which both diameter and contact angle are evolving with time. Sharp transitions from CCR to CCA (figure 1) are only observed for large volume drop. For volume drop less than 3 µl, a continuous transition from CCR to mixed regime is observed most of the time. As it can be seen on figure 3, the existence of pure CCR is also debatable and we are facing mixt regimes most of the time. For patterned surfaces (figure 2) we observed many small transition regimes, corresponding to local jump of the triple line. This is the stick-slip regime.

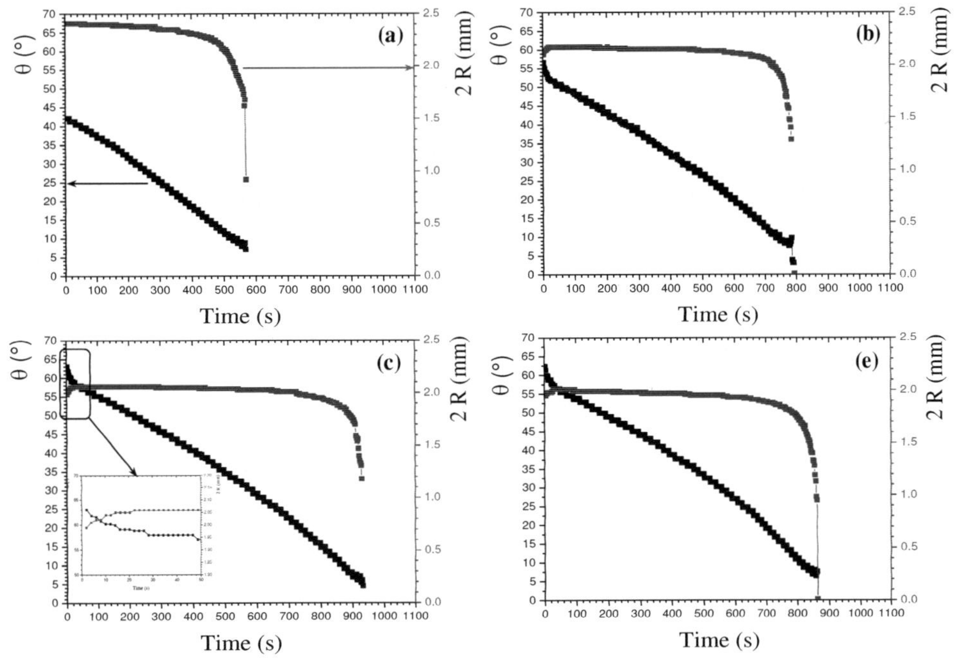

Figure 3. Evolution of the contact angle and drop width for drops of 1 µl as a function of time for different ageing times on polished stainless steel sample 2 exposed to air: (a) freshly polished, (b) 3 days, (c) 7 days and (d) 10 days of air exposure. The onset in (c) shows the early beginning of the drop evaporation.

IOP Conf. Series: Materials Science and Engineering **258** (2017) 012003 doi:10.1088/1757-899X/258/1/012003

3.2. Tracking the traces

Another important finding on polished stainless steel is the important aging effect on the contact angle. Kim et al [16] recently report on such an effect for polished and aged AISI STS 304 samples. Time-dependent wettability transitions were observed after laser [17-18] or electrochemical processing [19]. Oxygen, carbon dioxide, and organic matter in the atmosphere have been proposed to explain such ageing effect [16-18]. The contamination layer seems to dissolve in water producing tracks visible at the end of the evaporation (onsets in figure 1 and 2). Such a phenomenon must also be connected to the rapid decrease of the contact angle observed at the beginning of the evaporation process on contaminated surfaces (figure 3). We have to note that this phenomenon is not observable on freshly polished surfaces. Figure 4 shows the evaporation of a sessile droplet seen from the top by confocal microscopy.

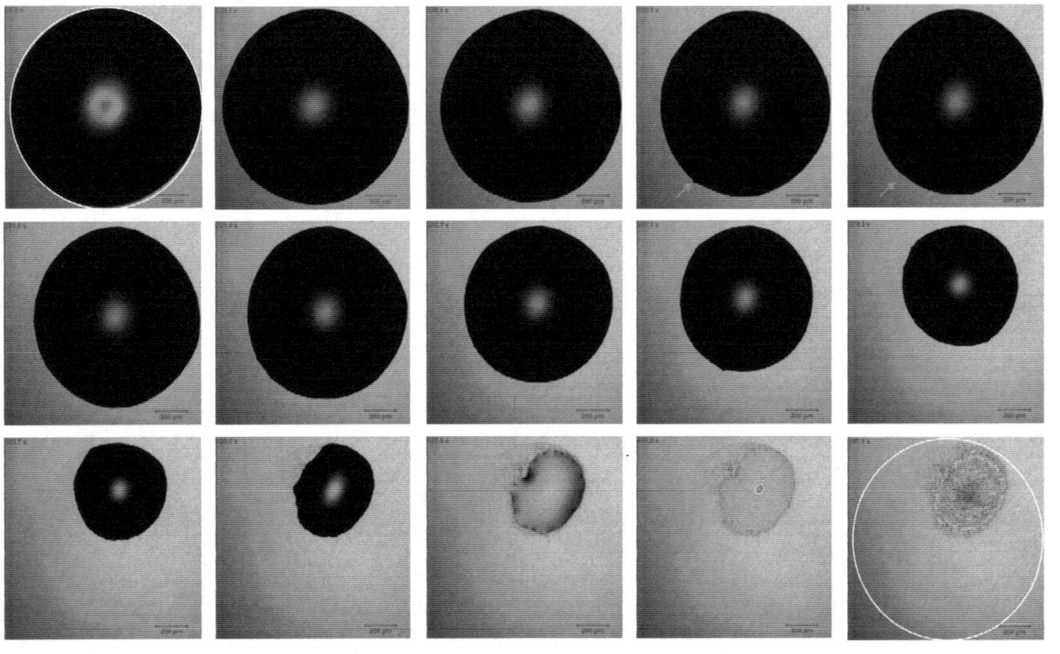

Figure 4. Evolution of the contact surface as a function of time for a 0.3 µl water drop on polished stainless steel sample 2 exposed to air for a long time as observed by confocal microscopy. Scale bar (200 µm)

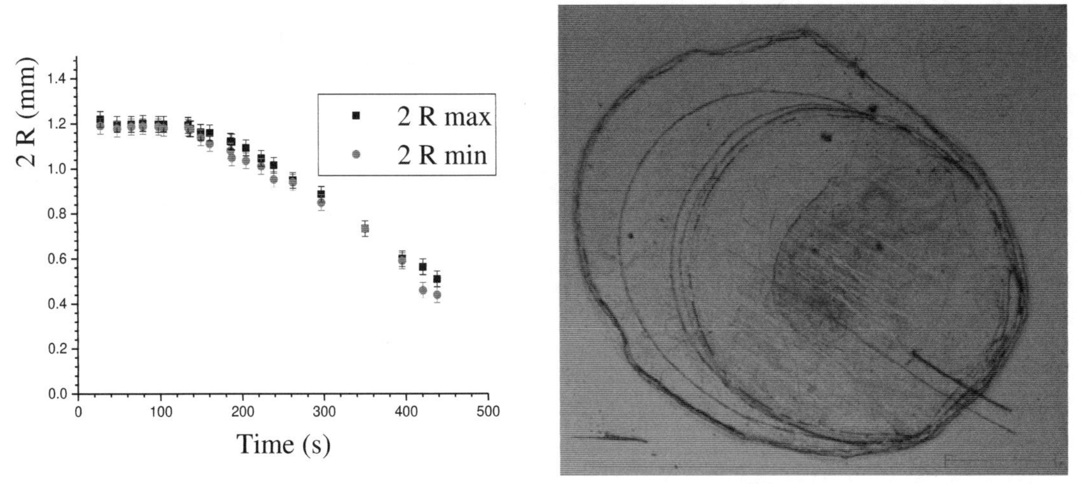

Figure 5. (a) Evolution of the drop width for drops of 0.3 µl as a function of time (data from fig.4), (b) traces remaining after the evaporation of a 3 mm in diameter water drop (approximately 4.2 µl). Both are for polished stainless steel samples 2 exposed to air for a long time

The triple line is almost circular during the evaporation process. The evolution of minimum and maximum radius of the triple line on the assumption of elliptical shapes are depicted on figure 5a. For such a small volume droplets (0.3 µl), the decrease of the diameter is more soft compared to larger drop volume (figures 1 and 3). Another interesting observation is in the kinetics of the triple line movement. It is not fully continuous, but instead is constituted by numerous stick slip micro-events (see arrows in figure 4). It is difficult to say to which type of discontinuities these events are associated. It can be due to dust or to other defects like grain boundaries. It can be noted that, on most of the observations made with the confocal microscope, a pinning of the triple line on the initial position of the perimeter is observed. This is illustrated by the white ellipse drawn on the last snapshot of figure 4, which allows a comparison of the trace of the evaporated droplet and its initial position. For small volumes of droplets, there is no multiple elliptical traces, as observed for larger volume (figure 5b). The mixt regime occurs quasi-continuously for small volumes, whereas sharp transitions from the CCR to CCA modes occurs for larger volumes, which can be seen by the multiple traces remaining on the surface at the end of the evaporation process (figure 5b).

3.3. Comparison between experimental and theoretical aspects for the evaporation of sessile droplets

The process of evaporation corresponds to the progressive transfer of a liquid into its own vapour. Such a mechanism is possible as long as the thermodynamic equilibrium is not established by the transition of the water molecules between the liquid and the gaseous phase. When the gaseous phase is saturated, there are as many incoming molecules as leaving the liquid phase and the equilibrium condition is realized. If the gaseous phase is not saturated, there are more molecules which pass into the gas phase and the evaporation phenomenon is present. The evaporation of a drop can be considered as a quasi-stationary phenomenon when the diffusion time is very small compared to the evaporation time. The concentration of vapour then readily adjusts with respect to the time required for evaporation. This is true when the density of the vapour at the surface of the drop is much lower than the density of the liquid (ratio of 10^{-5} for water). The mass flow of evaporation which is established on the surface of the drop is then modelled by the diffusion of molecules of the liquid-gas interface to the surrounding gas phase. When a drop evaporates in contact with a substrate, the problem is complex to solve because of the loss of symmetry and the appearance of a triple line. Droplets small enough to be associated with spherical caps (valid when the radius R of the tip is less than the capillary length: 2.7 mm for water) will be considered hereafter. Hence the volume variation is given by [20]:

$$\frac{dV}{dt} = \frac{\pi R^3}{(1+\cos\theta)^2}\frac{d\theta}{dt} + 3\pi R^2 g(\theta)\frac{dR}{dt} \text{ with } g(\theta) = \frac{\cos^3\theta - 3\cos\theta + 2}{3\sin^3\theta} \quad (1)$$

Popov [20] has shown that equation (1) can be given by:

$$\frac{dV}{dt} = -\pi R\,\frac{D(C_\infty - C_S)}{\rho} f(\theta) \text{ with } f(\theta) = \frac{\sin\theta}{1+\cos\theta} + 4\int_0^\infty \frac{1+\cosh 2\theta\tau}{\sinh 2\pi\tau}\tanh[\tau(\pi-\theta)]\,d\tau \quad (2)$$

ρ is the fluid density, $f(\theta)$ is a function of the contact angle θ, D is the diffusion constant for vapour in air, C_S is the vapour saturation concentration just above the liquid-air interface, C_∞ is the ambient vapour concentration. This last relation is relevant for all the θ values. When $\theta \to 0$, relation (2) gives [20]:

$$\theta = \theta_i\left(1-\frac{t}{t_f}\right) \text{ and } V = \frac{\pi R_i^3\theta_i}{4}\left(1-\frac{t}{t_f}\right) \quad (3)$$

$$t_f^{CCR} = \frac{\pi\rho R_i^2\theta_i}{16 D(C_\infty - C_S)} \quad (4)$$

In relations (3) and (4), θ_i is the initial contact angle and t_f^{CCR} is the evaporation time in the CCR regime. Relation (2) has no simple analytical solutions but some analytical expression have been proposed for $f(\theta)$ [21]. In the CCR mode, the obtained behaviours for θ and V are similar to those given by relations (3) and (4). In the CCA mode, relations (5) and (6) are obtained:

$$R = \sqrt{R_0^2 - \psi(\theta_0)\,t} \text{ and } V = V_0 - 2\pi\left(\sqrt{R_0^2 - \psi(\theta_0)}\right)\left(\frac{D(C_\infty - C_S)}{\rho}f(\theta_0)\right)t^{\frac{3}{2}} \quad (5)$$

$$t_f^{CCA} = \frac{R_0^2}{\psi(\theta_0)} \text{ with } \psi(\theta_0) = \frac{4}{3}\frac{D(C_\infty - C_S)}{\rho}\frac{f(\theta_0)}{g(\theta_0)} \quad (6)$$

In relation (5) and (6), θ_0 is the contact angle at the beginning of the CCA mode and t_f^{CCA} is the evaporation time in the CCA regime. As a general rule, the simple evaporation model gives the following power law for the droplet volume evolution as a function of time (V_i is the initial volume of the droplet):

$$\left(\frac{V}{V_i}\right)^\beta = \left(1-\frac{t}{t_f}\right) \quad (7)$$

Works from Nguyen et al [21] and Hu et al [22] show that the β factor is evolving between 1 and 2/3 for the CCR mode and is equal 2/3 for the CCA mode.

Figure 6a and 6b respectively shows the evolution of the logarithm of the normalized drop volume as a function of ln $(1- t/t_f)$ for the data of figures 1 and 2. During most of the evaporation process, the evolution of ln V_{norm} as a function of ln $(1- t/t_f)$ is linear for both the polished and the patterned stainless steel samples (with respective slops of 1.08 and 0.9). A slope of 0.8 was found for smaller droplet volumes (data of figure 3). It can be concluded that relation (3) for the evaporation in the CCR mode is effective in both cases. However, the final part of the evaporation process is not correctly described by relation (3) as noted by Popov [20]. We also note that the theoretical value for the CCA

IOP Conf. Series: Materials Science and Engineering **258** (2017) 012003 doi:10.1088/1757-899X/258/1/012003

mode (3/2) is not observed for the end of the evaporation process (0.2 for the patterned sample). This is illustrated in the evolution of the droplet volume as a function of the evaporation time represented in figure 6c. If the initial volume evolution is correctly described by a linear law, it is not the case for the end of the evaporation process. From figure 6c, it can be also shown that the decrease of both contact angle and drop radius observed in the CCR mode of figure 2 (between 240 and 280 s) has not physical signification. As a matter of fact, during this time, the temporary decrease in volume of the droplet observed is followed by an increase in the droplet volume correctly adjusted to the linear evolution as a function of time. This event is probably connected to a temporary non symmetric changes in the shape of the sessile droplet that cannot be correctly recorded by our experimental device.

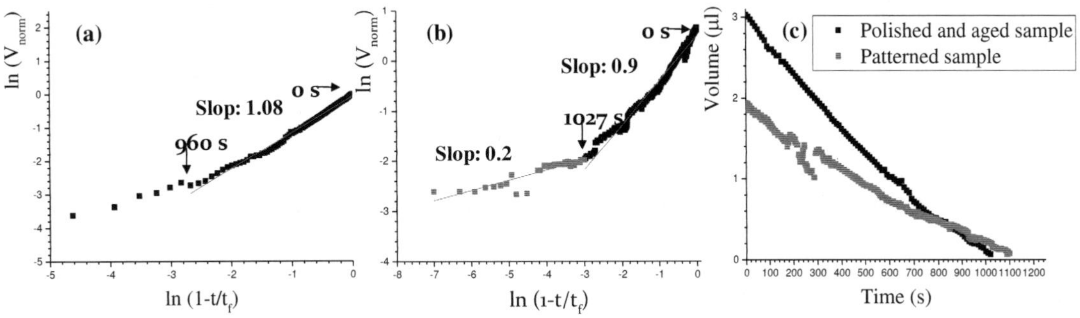

Figure 6. Evolution of the logarithm of the normalized droplet volume as function of the logarithm of (1- t/t_f) for (a) data of figure 1, (b) data of figure 2. (c) Evolution of the volume as function of time for the corresponding two cases.

The evolution of the volume of the drops is correctly described by the theory of evaporation in CCR mode, what is about the evaporation time t_f? From relation (4) it should depend on $(R_i)^2\,\theta_i$ in the CCR mode. Figures 7a and 7b respectively show the evolution of the initial droplet radius (R_i) and of the initial contact angle (θ_i) as a function of the evaporation time t_f for different polished and aged stainless steel samples and for the data of figures 1 and 2. For these data, different drop volumes are taken into account (1 µl for the polished and aged samples of figure 3, 1.9 µl for the patterned sample from figure 2 and 3 µl for the polished and aged sample from figure 1). From figure 7a, it is clear that the evaporation time does not strongly depends on the initial radius of the droplets. However, the evaporation time is linearly depend on the evaporation angle as it can be observed from figure 7b. Such an observation, which is very important for the improvement of the evaporation processes at industrial scale, has no correct interpretation for the moment. Such a dependence of t_f on θ_I will be interesting to consider in order to improve the modelling of the evaporation process.

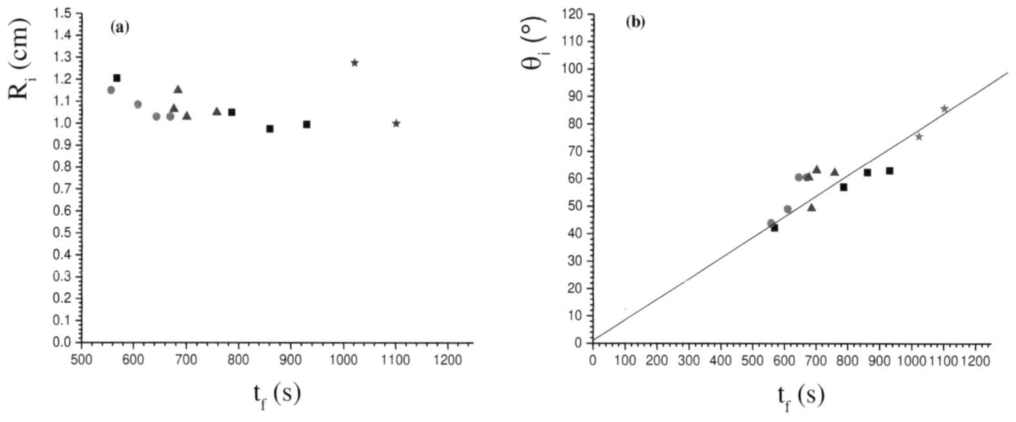

Figure 7. Evolution of (a) the initial droplet radius (R_i) and (b) of the initial contact angle (θ_i) as a function of the evaporation time (t_f) for different polished and aged samples and for the data of figures 1 and 2.

4. Conclusion

Evaporation of sessile water droplets was studied on different surface states of austenitic stainless steel: mirror polished, mirror polished and aged and patterned by sputtering. The evolution of the contact angle and of the droplet diameter was presented as a function of time at room temperature. For all the surface states, a constant contact radius evaporation mode (CCR) was observed. On polished and aged surfaces, it was sometimes followed by a nearly constant contact angle regime (CCA) for large volume droplets. In most case, a mixed regime, in which both the contact angle θ and the radius R of the drop decreases with time, was observed. For the patterned surface and for small droplet volumes a stick-slip mode was recorded. An important aging effect on the contact angle was measured on polished surfaces. It was attributed to atmospheric contamination. Confocal microscopy observations from the top of the drop have shown that the triple line is almost circular during the evaporation process. A pinning of the triple line on the initial position of the perimeter was observed for polished and aged surfaces.
The experimental observations were compared to a quasi-static evaporation model for which the droplets are small enough to be associated with spherical caps. In our experiments, the evolution of the droplet volume as a function of time is almost linear with the evaporation time for all the observed surfaces. This is in accordance with the model prediction for the CCR mode for small initial contact angles (θ_i). However, the final part of the evaporation curves is not correctly described by the model. Also, the evaporation time (t_f) is not correctly described by the evaporation model which involves a R_i^2 θ_i dependence for t_f, whereas a θ_i dependence is observed for the studied surfaces. It will be interesting to consider this dependence in order to improve the modelling of the evaporation process.

5. References

[1] Quéré D, 2013 Annu. Rev. Fluid Mech., **45** 197
[2] Rein M, 2002 *Drop surface interaction*, (Springer-Verlag Wien GmbH, M. Rein, ed.)
[3] Liang G and Mudawar I, 2017 Int. J. of Heat and Mass Transfer **106**103
[4] Duursma G, Kennedy R, Sefiane K, Yu Y, 2016 Heat Transfer Engi., **37** 1190.
[5] Marcos G, Guilet S, Cleymand F, Thiriet T, Czerwiec T, 2011 Surf. Coat. Technol. **205** S275
[6] Czerwiec T, Marcos G, Thiriet T, Guo Y, Belmonte T, 2009 IOP Conf. Series: Materials Science and Engineering **5** 012012.

[7] Bhushan B and Jung Y C, 2010 Progress in material Science **5** 1.
[8] Kreder M J, Alvarenga J, Kim P, Aizenberg J, 2016, Nature Reviews Materials, **1** 1.
[9] Ueda E and Levkin P A, 2013 Adv. Mater., **25** 1234.
[10] Tao P , Shang W , Song C, Shen Q, Zhang F, Luo Z , Yi N , Zhang D, Deng T, 2015 Adv. Mater., **27** 428
[11] Zhang P and Lv F Y, 2015 Energy **82** 1068
[12] Stauber J M, Wilson S K, Duffy B R, Sefiane K, 2014 J. Fluid. Mech. **744** R2.
[13] Gelderblom H, Marin A G, Nair H, Van Houselt A, Lefferts L, Snoeijer J H, Lohse D, 2011 Phys. Rev. E **83** 026306.
[14] Bormashenko E, Musin A, Whyman G, Zinigrad M, 2012 Langmuir **28** 3460.
[15] Semenov S, Trybala A, Agogo H, Kovalchuk N, Ortega F, Rubio R G, StarovV M, Velarde M G, 2013 Langmuir **29** 10028
[16] Kim D, Kim J, Chu C N, 2016 Materials Letters **170** 18
[17] Kietzig A M, Hatzikiriakos S G, Englezos P, 2009 Langmuir, **25** 4821.
[18] Kruse C, Anderson T, Wilson C, Zuhlke C, Alexander D, Gogos G, Ndao S, (2013) Langmuir **29** 9798.
[19] Liu P, Cao L, Zhao W, Xia Y, Huang W, Li Z 2015 Appl.Surf.Sci. **324** 576
[20] Popov Y O, 2005 Phys. Rev. E **71** 036313.
[21] Nguyen T A H, Nguyen A V, 2014 Langmuir, **30** 6544.
[22] Hu D, Wu H, 2016 Phys. Rev. E, **93** 042805.

Acknowledgments
This work and the post-doctoral stay of S. Tsareva were supported by the French Government through the programs Institut Carnot ICÉEL, and "Investissements d'avenir" operated by the French National Research Agency (ANR) and referenced to as ANR-11-LABX-0008-01 (LabEx DAMAS).

IOP Conf. Series: Materials Science and Engineering **258** (2017) 012004 doi:10.1088/1757-899X/258/1/012004

Combination of experimental and numerical methods for mechanical characterization of Al-Si alloys

A Kruglova[1], M Roland[2*], S Diebels[2], F Mücklich[1]

[1] Chair of Functional Materials, Saarland University, D-66123 Saarbrücken, Germany
[2] Chair of Applied Mechanics, Saarland University, D-66123 Saarbrücken, Germany

* m.roland@mx.uni-saarland.de

Abstract. In general, mechanical properties of Al-Si alloys strongly depend on the morphology and arrangement of microconstituents, such as primary aluminium dendrites, silicon particles, etc. Therefore, a detailed characterization of morphological and mechanical properties of the alloys is necessary to better understand the relations between the underlined properties and to tailor the material's microstructure to the specific application needs. The mechanical characterization usually implies numerical simulations and mechanical tests, which allow to investigate the influence of different microstructural aspects on different scales. In this study, the uniaxial tension and compression tests have been carried out on Al-Si alloys having different microstructures. The mechanical behavior of the alloys has been interpreted with respect to the morphology of the microconstituents and has been correlated with the results of numerical simulations. The advantages and limitations of the experimental and numerical methods have been disclosed and the importance of combining both techniques for the interpretation of the mechanical behavior of Al-Si alloys has been shown. Thereby, it has been suggested that the density of Si particles and the size of Al dendrites are more important for the strengthening of the alloys than the size-shape features of the eutectic Si induced by the modification.

1. Introduction

Al-Si alloys are often the material of choice for many automotive applications such as cylinder heads [1] and engine blocks [2] due to their attractive strength characteristics combined with the low weight of casting components. In this respect, mechanical properties of the alloys play a principal role in assessing their functionality. The mechanical behavior of Al-Si alloys has a morphological genesis, i.e. it is defined by the morphology and arrangement of the microconstituents [3], such as primary aluminium dendrites, silicon particles, intermetallic inclusions and casting defects. Therefore, a thorough characterization of morphological and mechanical properties of the alloys is necessary to better understand the relations between the underlined properties of the material and to tailor its microstructure to the specific application requirements.

Regarding the morphological impact of Si, the connectivity of Si particles has been found to strongly influence the mechanical behavior of Al-Si alloys [4, 5]. In particular, it has been shown that the strength of the eutectic structure increases with the connectivity and the branching of Si particles. Besides, the higher the number of Si particles in disjoint or simply connected structures, the higher the strength of the corresponding structures [6]. While some researchers suggest that the ductility of the alloys depends on the morphology of both the primary and the eutectic phase [7–11], others attribute a primary role in mechanical properties either to Si particles [3, 12] or to Al dendrites [13]. In general, the ductility of Al-Si alloys decreases with increasing secondary dendrite arm spacing (DAS) [14], with the exception of alloys with large DAS values [8–10].

Different methods of mechanical characterization, such as numerical simulations and mechanical tests, can be used to study the influence of the microstructure on the mechanical behavior of the material. For example, the effect of Si connectivity can only be evaluated in a volume [15], i.e. via finite elements method (FEM) simulations carried out on 3D image data representing geometrical properties of microconstituents. The mechanical behavior obtained thereby can shed light on the local stress-strain state in a volume but is not representative of the entire material since it is limited to the simulation volume solely. On the contrary, tension and compression tests conducted on specimens of casted alloys deliver results that are relevant for the interpretation of the material's behavior on the macro-scale. Still, the impact of different microconstituents cannot be assessed independently because the material consists of several phases whose amounts and morphology are difficult to control during the manufacturing process.

In this study, the uniaxial tension and compression tests have been carried out on different Al-Si alloys. The mechanical behavior of the alloys has been interpreted with respect to the morphology of the microconstituents and has been correlated with the results of FEM simulations performed elsewhere by the authors [6, 16]. Thereby, the advantages and limitations of the experimental and numerical methods of mechanical characterization of Al-Si alloys have been exposed. The aim of the present investigation has been to show how the combination of both methods can contribute to a better understanding and interpretation of the mechanical behavior of the material observed on the macroscopic level with respect to the morphological properties of its microconstituents, in particular, primary aluminium dendrites and eutectic phase.

2. Materials and methods

2.1. Materials

The Al-Si alloys have been elaborated by Hydro Aluminium Rolled Products GmbH by means of the permanent mold casting. Two types of alloy compositions have been investigated: an unmodified AlSi12 and an AlSi12 modified with 212 ppm of Sr, also referenced as 2X and 3X, respectively.

The alloys have been investigated in the as-cast (AC) and solution treated (ST) state. Specimens machined for the compression and tensile tests have been heated in a High Temperature Chamber Furnace type RHF 1600 of Carbolite Company. The heat treatment cycle has been composed of the heating up to 540° C, solution treatment for 4 hours and quenching in water near the boiling point to lower the level of induced residual stresses [17].

2.2. Compression tests

The compression tests have been carried out with the help of the E10000 Linear-Torsion Floor Instrument of Instron® Company, which allows a maximum load capacity of 10 kN. All test specimens have been cut from similar parts of the castings into cubes with the side length of 5 mm. The tests have been performed at a constant cross-head speed of 0.001 mm/s.

2.3. Tensile tests

The tensile tests have been carried out with the same machine as the compression tests. The specimens have been machined according to the European Standard EN ISO 527-2 [18], where the gage has a square cross-section of 2×2 mm^2 and is 10 mm long. The cross-head speed of 0.01 mm/s that corresponds to the strain rate of 1×10^{-3} s^{-1} has been used for the tensile testing procedure. The engineering stress-strain curves have been computed by dividing the force by the initial cross-section area of the specimen and the amount of stretch of the gage by the initial gage length of the specimen, respectively.

IOP Conf. Series: Materials Science and Engineering **258** (2017) 012004 doi:10.1088/1757-899X/258/1/012004

2.4. FEM simulations
The FEM simulations have been carried out on 3D data of Al-Si eutectic phase reconstructed via the FIB/SEM (focused ion beam/scanning electron microscope) tomography. For more details on the imaging and data processing techniques, the reader is referred to [16]. The reconstructed and segmented 3D image data have been meshed with tetrahedral elements, and the material properties have been mapped directly into each mesh cell. The following properties have been assigned to the materials: A Young's modulus of $E = 70$ GPa, a Poisson's ratio of $v = 0.34$, and a yield strength of $\sigma_y = 40$ MPa have been applied to the aluminium. A Young's modulus of $E = 107$ GPa, a Poisson's ratio of $v = 0.27$, and a yield strength of $\sigma_y = 7$ GPa have been assigned to the silicon. All simulations have been executed with the finite element software Abaqus using linear finite elements and the explicit solver, with a load curve reproducing the conditions of the compression tests (20 % deformation).

3. Results and discussion

3.1. Micrography
The microstructure of the alloys consists of primary Al dendrites and Al-Si eutectic. Although the alloys have the eutectic composition with a Si content of 12 wt.%, in some samples the volume fraction of Al dendrites reaches the amount of nearly 30%. Normally, the modification of hypoeutectic Al-Si alloys results in shifting the eutectic point towards a higher Si level and as a consequence, a higher amount of the primary phase in the solidified microstructure [19–21]. Liao et al. [22] have observed that the amount of the primary aluminium in the near-eutectic Al-Si alloys increases with increasing Sr content, in particular, when the amount of Sr is above 0.015%. However, this effect of the modification has not been observed in the investigated alloys. On the contrary, no significant difference in the primary phase content between the modified and unmodified alloys has been detected in the present investigation: The amount of Al dendrites in the modified alloy with 0.02% of Sr is even slightly lower than in the unmodified one (26% versus 28%, respectively), although they are both higher than the equilibrium level. Therefore, it can be concluded that a higher amount of the primary phase with respect to the equilibrium level is related to the high cooling rate during the solidification. This result is in accordance with the observations reported in [21]. Thus, it is likely that the effect of the high cooling rate dominates the effect of the modification, at least, in the present investigation.
The optical micrographs in figure 1 show the (a) unmodified and (b) Sr-modified AlSi12 alloys, particularly zooming into the eutectic phase. The modified structure is very fine and homogeneous: Si particles have mostly spherical cross sections that are simply the cuts through fibrous or coral-like particles in 3D [23]. The unmodified alloy exhibits less homogeneous Si profiles ranging from spherical through elongated profiles to profiles of irregular shape. Such cross sections belong to the plate-like particles that are typical of the unmodified Al-Si alloys [23]. Intermetallic phases are not observed. Although the unmodified Si particles are normally coarser than those of the modified alloy, both investigated structures are relatively fine because of the small dimensions of the original ingots and high cooling rates during the casting in the permanent mold.

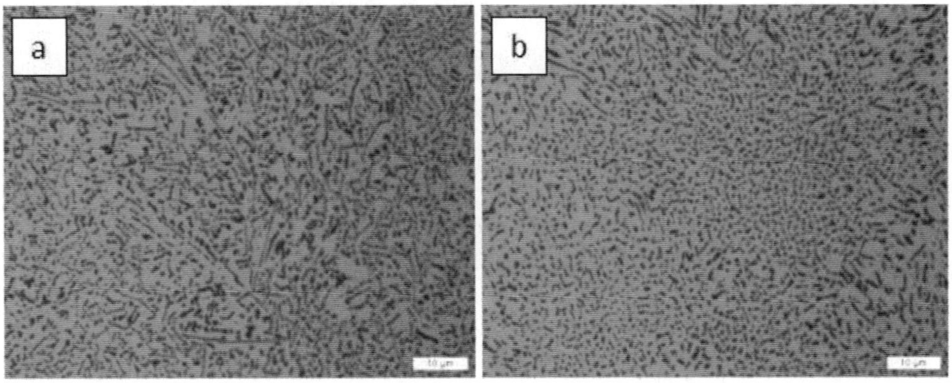

Figure 1. Optical micrographs of the (a) unmodified AlSi12 and (b) Sr-modified AlSi12, zoom in the eutectic region: the eutectic consists of dark-grey Si particles encompassed in the light Al matrix.

The optical micrographs in figure 2 show the (a) unmodified and (b) Sr-modified AlSi12 alloys after the solution treatment. Both structures have undergone a considerable agglomeration and spheroidization of Si particles.

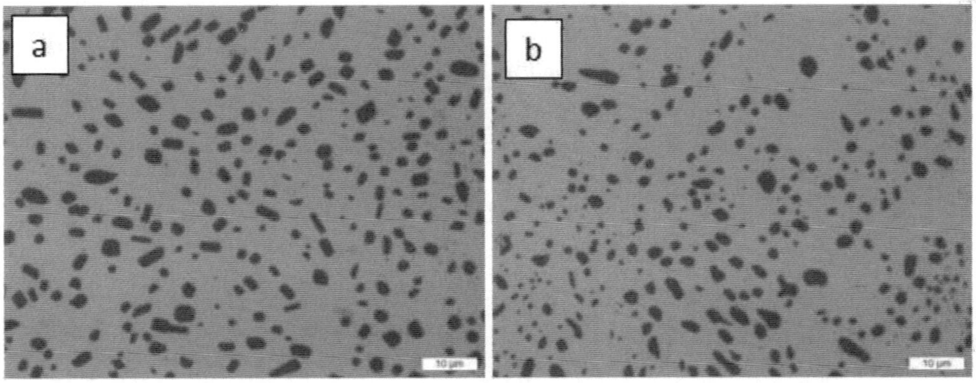

Figure 2. Optical micrographs of the (a) unmodified AlSi12 and (b) Sr-modified AlSi12, both are solution heat treated at 540° C for 4 hours and quenched in boiling water. In both alloys Si particles have been agglomerated and spheroidized. The particles are much coarser in comparison to the corresponding structures of the as-cast alloys.

As reported in [24], Sr modification influences the kinetics of Si spheroidization and coarsening during the solution heat treatment in such a way that modified Si particles have more propensity for the spheroidization due to their initial smooth fibrous shape. On the contrary, unmodified Si plates are more resistant to shape changes and thus, rather undergo coarsening and, to a lesser extent, spheroidization. The authors also distinguish a shape-size diversity in unmodified alloys to be an additional driving force for the coarsening. However, both structures in figure 2 do not exhibit any pronounced difference in shape and/or size of the particles. Even if any difference did exist at a first stage of the temper, it should have been diminished after 4 hours, since, on the one hand, both alloys have a priori shown a similar fine eutectic structure, and on the other hand, the solution treatment time has been too extended for such small samples. Thus, a similar mechanical behavior of the alloys after the solution treatment can be expected.

3.2. Mechanical behavior

The mechanical behavior of the as-cast and solution treated Al-Si alloys has been investigated under compression and tensile loading. The corresponding stress-strain curves are shown in figure 3 and figure 4, respectively.

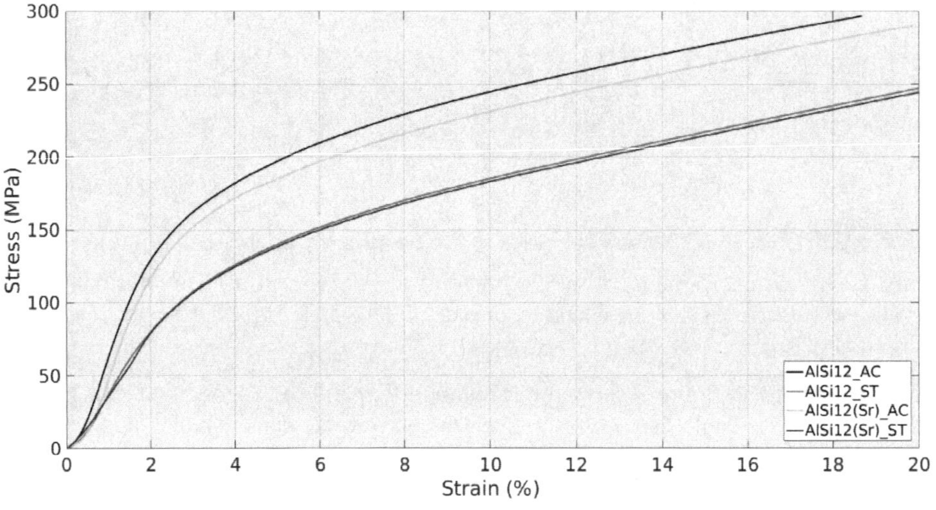

Figure 3. Engineering stress-strain curves obtained from the compression testing of AlSi12 (2X) and AlSi12(Sr) (3X) alloys.

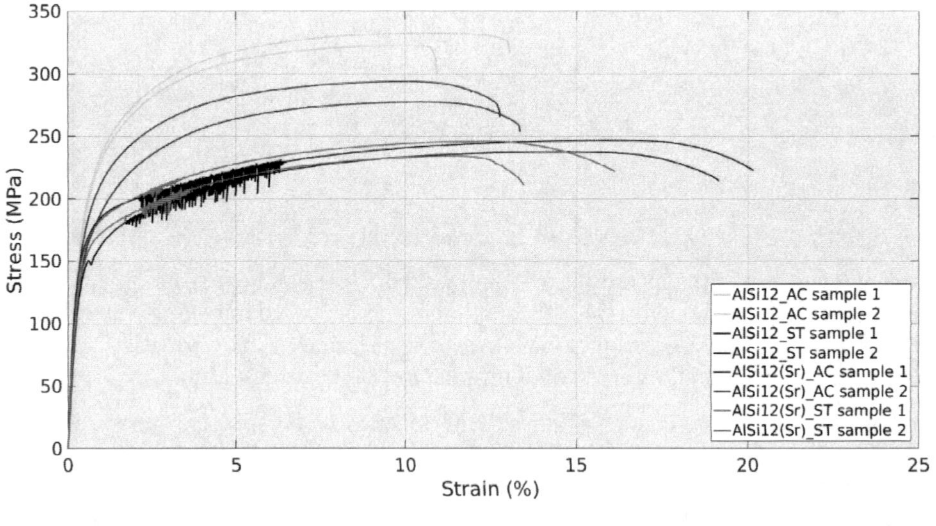

Figure 4. Engineering stress-strain curves obtained from the tensile testing of AlSi12 (2X) and AlSi12(Sr) (3X) alloys. The oscillations on the curves of solution treated alloys, 2X and 3X, are due to the calibration of the testing machine to a softer material, and can therefore be neglected.

As can be seen in figure 3 and figure 4, the as-cast samples exhibit a higher strength than the solution treated samples of the corresponding alloys. During the solution treatment, Si particles undergo disintegration and spheroidization, which lead to an unavoidable reduction of the load carrying capacity of the structure [4]. According to Requena et al. [5], the stiff Si network in the ductile Al matrix increases the load transfer from the matrix to the reinforcement, which improves the strength.

Therefore, the disintegration of the network of Si particles and as a consequence, the reduction of their connectivity induced by the heat treatment has an adverse effect on the strength of Al-Si alloys, although reducing the strength it improves their ductility.
The experimental results are also consistent with the observations on the influence of the Si connectivity on the compressive strength of Al-Si alloys reported in [6]. In contrary to the partially interconnected Si particles in the as-cast alloys, Si particles in the solution-treated alloys are simply connected. Thus, the load transfer from the matrix to these particles is lower than the one to the Si particles with a higher connectivity. Hence, the strengthening effect of the spheroidized Si in comparison to the more complex-shaped particles is also reduced. The reduction of the load-bearing capacity of spherical particles in relation to the high aspect ratio particles (e.g. fibrous or plate-like particles) has also been discussed in [25–27].
Besides, the ductility of the solution-treated samples is higher than the ductility of the as-cast alloys. Similar observations have been reported by Guiglionda and Poole [26], who have examined an impact of the heat treatment on Si morphology and ductility of Al-Si eutectic alloys. In general terms, the fragmentation of Si particles with the subsequent spheroidization during the solution heat treatment is beneficial for the ductility of the material [9, 24].
There are two more observations that can be made in figure 4. First, in contrary to the commonly accepted superiority of the strength characteristics and ductility of modified alloys as compared to unmodified ones, the unmodified alloy, 2X_AC, exhibits a higher strength than its modified counterpart, 3X_AC. The second observation is that in the ST condition, this difference disappears and the alloys show similar strength characteristics. Although it has been expected that the alloys with a similar microstructure (see figure 2) have a similar mechanical behavior, the alloy 2X_ST shows a higher elongation to fracture than the alloy 3X_ST. Furthermore, the stress-strain curves of both alloys show strong oscillations that are most probably due to the calibration of the testing machine to a softer material, and can therefore be neglected. To exclude the possibility that such a behavior is observed due to statistical issues, complementary tensile tests have been carried out. The results of the tensile tests performed on the modified and unmodified Al-Si alloys in the ST and AC conditions are shown in figure 5 and figure 6, respectively.

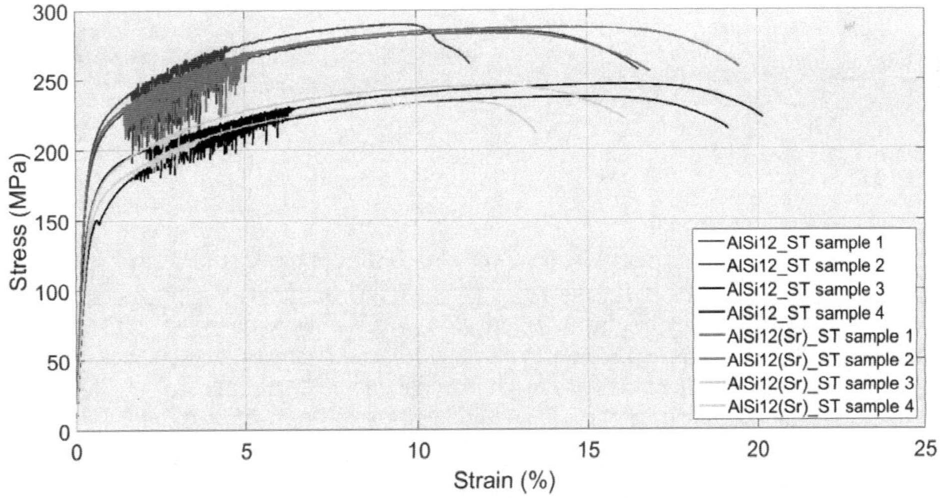

Figure 5. Engineering stress-strain curves obtained from the tensile testing of the unmodified (2X_ST) and Sr-modified (3X_ST) Al-Si alloys. The complementary tests have been carried out on samples 1 and 2 showing higher strength characteristics than the original samples 3 and 4 for both alloys. The oscillations are still present on the curves.

As can be seen in figure 5, the complementary tests (samples 1 and 2) have yielded higher strength characteristics than the original tests (samples 3 and 4). However, the oscillations are still present on the curves but in contrary to the previous results, the stress-strain curves from the complementary tests show that the alloy 3X_ST has a higher elongation to fracture than the alloy 2X_ST. Yet, as expected, the alloys exhibit similar average strength characteristics and elongation to fracture. The quantitative data on the average mechanical properties of the investigated alloys derived from the stress-strain curved are shown in table 1.

The results deduced from the stress-strain curves include the following mechanical characteristics: yield strength, ultimate tensile strength, and total elongation. Since Al-Si alloys do not exhibit a pronounced yield point, an offset deformation of 0.2% has been used to define the yield strength [28]. The corresponding structural and mechanical properties of the analyzed Al-Si alloys are summarized in table 1.

Table 1. Mechanical properties of the analyzed Al-Si alloys represented by yield strength, *YS*, ultimate tensile strength, *UTS*, and elongation, *e*, and size of Si particles represented by equivalent particle diameter, *d*, in the corresponding alloys.

Alloy	d (μm)	YS (MPa)	UTS (MPa)	e (%)
2X_AC	0.89	195	338	12
2X _ST	2.02	176	265	17
3X_AC	0.78	184	323	13
3X_ST	1.53	177	263	16

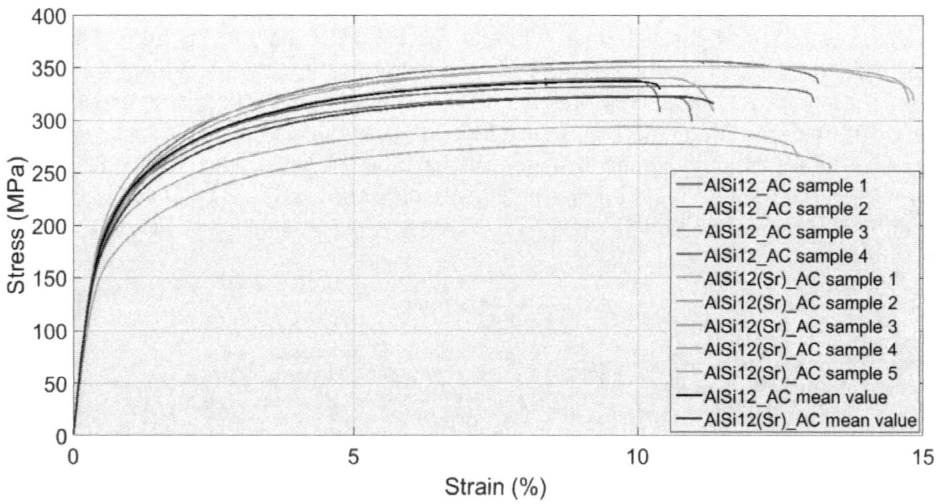

Figure 6. Engineering stress-strain curves obtained from the tensile testing of the unmodified (2X_AC) and Sr-modified (3X_AC) Al-Si alloys. Sample 5 of the unmodified AlSi12 alloy has been excluded on purpose due to its failure at the very beginning of the experiment and hence, unrepresentative behavior. On the average, the alloy 2X_AC has a higher tensile strength than the alloy 3X_AC. Please note that the average stress-strain curves are computed for the stress values only.

Both the unmodified (2X_AC) and modified (3X_AC) Al-Si alloys have a similarly fine structure of the eutectic Si (see figure 1). Therefore, as has been expected, both alloys also have a similar mechanical behavior under compression and tension loading, though with a slight strength superiority of the alloy 2X_AC. As can be clearly seen in figure 6, on the average, the strength of the unmodified alloy is higher than the strength of the modified alloy. The same observations have been made by analyzing the mechanical behavior of the alloys under compression loading (see figure 3). However, metallographic observations of the eutectic Si morphology as well as the state-of-the-art in the field of Sr modification of Al-Si alloys [19, 23, 24, 29–31] argue for the expectations of the opposite behavior.

According to Hafiz and Kobayashi [3], more fine and regular modified particles have less stress concentration sites and can carry the load much better due to the more gradual damage evolution. It can therefore be suggested that the clue to the understanding of the observed phenomena lies in the three-dimensional nature of the Si structure and/or properties of the primary aluminium phase.

Considering the influence of the primary aluminium phase, it has to be taken into account that the macro samples of Al-Si alloys used for the mechanical tests do not consist of the eutectic phase only but contain up to 30% of Al dendrites. Some researchers [7–11] suggest that both microconstituents, Al dendrites and Si particles, play an important role for the tensile properties and fracture behavior of the alloys. Therefore, the dendrite arm spacing (DAS) characterizing the size of the primary aluminium phase has been determined. It has turned out that DAS in the unmodified alloy (2X_AC) is lower than the one in the modified alloy (3X_AC) and is equal to 14.08 ± 2.01 μm and 19.28 ± 3.07 μm, respectively. In general, the refinement of the primary phase corresponds to the improvement of tensile properties [11], which agrees well with the experimental observations.

To analyze the impact of the three-dimensional structure of Si particles, the connectivity of the eutectic Si in both alloys has to be determined. Here, the connectivity of the eutectic Si is assessed with the help of the Euler number (Euler number density) [15, 32] that represents the difference between the number of individual Si particles and the connectivity and can only be determined in a volume via different image analysis software. To do so, the 3D morphology of the eutectic phase has been reconstructed after the mechanical tests via the FIB/SEM tomography. To measure the initial (prior to deformation) connectivity of Si in the investigated alloys, the cracks in Si particles induced by the mechanical loading have been filled using image editing methods. In doing so, the computed connectivity could represent the connectivity of Si particles in the undamaged state [33]. Furthermore, to compute the Euler number density of Si particles in the undeformed samples, the estimated Euler number has to be divided by the initial (undeformed) volume of the reconstructed samples. By deriving the relation between the initial volume, V_0, the deformed volume, V, where $V = V_0 + \Delta V$, and the strain level, $\Delta L_l/L_0$, of the samples given the Poisson ratio of Al-Si alloys, ν, equation (1) has been obtained:

$$\frac{\Delta V}{V_0} \approx (1 - 2\nu)\frac{\Delta L_l}{L_0} \qquad (1)$$

After having computed the initial volume of the samples, V_0, using equation (1), the Euler number density has then been determined. According to the results shown by the authors in [6], a positive value of the Euler number density of 4.41 × 10^{17} m^{-3} in the eutectic phase of the sample 2X_AC against 2.04 × 10^{16} m^{-3} in the eutectic phase of the sample 3X_AC indicates a higher density of Si particle clusters in the unmodified alloy, and as a consequence, a higher strength of the material. This result also agrees well with the experimental observations.

Thus, it can be suggested that the strength characteristics observed in the modified and unmodified alloys have resulted from the combined effect of the eutectic Si morphology and the size of the primary aluminium dendrites. In general, the microstructure refinement, which implies the refinement of both phases, is beneficial for the strength [34, 35] and ductility [9] of Al-Si alloys. In particular, the reduction of DAS results in the improvement of the tensile strength and ductility [30]. A higher number of particles in simply connected structures also improves the strength properties of the material [6]. These conclusions are supported by the experimental observations that the unmodified alloy with a lower DAS and a higher density of Si particles in the eutectic volume has a slightly higher strength than the modified alloy. At the same time, the alloys exhibit similar values of the average elongation to fracture (see table 1). While the (slightly) finer and more homogeneous Si structure in the modified alloy should normally lead to a moderate increase of the strength in comparison to the unmodified alloy, the higher DAS values and lower particle density exert an opposite effect. As a result, the density of the second phase particles and the size of the primary aluminium phase have proven to be more important for the strengthening of the alloy than the size-shape features of the eutectic Si induced by the modification. When the microstructure is fine, the advantages of the modified Si morphology for the ductility of the material might be overridden by the properties of the primary phase.

3.3. Stress-strain state in simulation volume

The stress state within the Al-Si eutectic resulting from the FEM simulations of uniaxial compression is shown in figure 7. It can be seen that some geometrical configurations in Si particles are more prone to a high stress concentration than others. In particular, the increased stress concentration is mostly observed in necking areas and branching points within the particles.

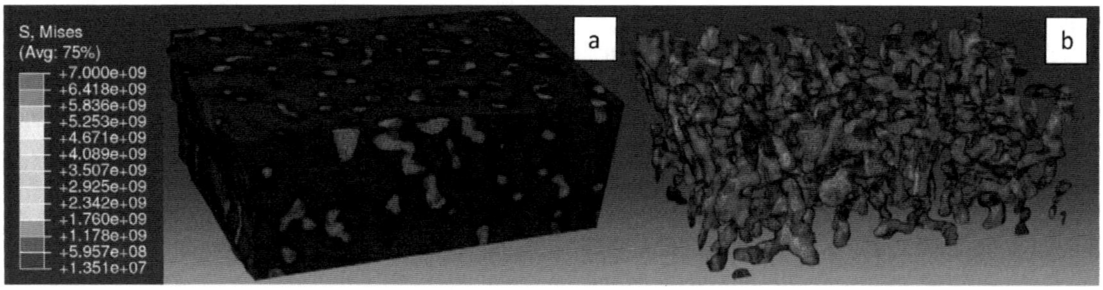

Figure 7. The stress state within the (a) Al-Si eutectic and (b) Si particles resulted from the FEM simulations of uniaxial compression along *y*-axis. The size of the original simulation volume is 36.6 μm × 16.8 μm × 34.9 μm. For better visualization, the simulation volume has been cropped in the middle along *x*-axis.

Evidently, the state of stress in Si particles and their fracture susceptibility in different morphological scenarios will vary due to the different load transfer from the matrix to Si particles, which depends on the particle size, shape, orientation and spatial arrangement (incl. the connectivity). FEM simulations have been shown to be highly effective in characterizing the mechanical behavior of Al-Si eutectic in 3D [6, 16, 36], and thus complementing the results of FEM simulations performed using a 2D multi-particle model [37] and microstructure based model [27] of Al-Si eutectic phase. 3D FEM simulations have allowed to investigate the influence of the connectivity and branching of Si particles [6], which is not accessible using 2D characterization techniques. Thus, 3D simulations deliver more accurate predictions of the stress state as compared to 2D simulations since they take into account 3D complexity of the structure. On the other hand, 2D FEM simulations have enabled to study the impact of morphological properties of Si particles (size, shape and orientation) on their stress state and consequently, fracture susceptibility in simply connected structures. In particular, the highest stress level has been detected in the particles oriented at 0° and 90° to the compression loading axis and exhibiting a higher aspect ratio [37], where the highest stress concentration within an individual Si particle has been found at sharp edges and bent regions [27]. However, one has to bear in mind that although 2D FEM simulations are in general consistent with experimental observations, they have a limited predictive power in relation to complex irregular or network structures. Moreover, the discussed 2D and 3D simulations do not take into account the residual stresses in the particles and their failure criteria [27].

It has also been observed from the results of FEM simulations [6], that an onset of the plastic deformation in the material occurs at much smaller strain levels than during the real mechanical tests. This is related to the inhomogeneous distribution of strain in the Al matrix within a simulation volume, where certain regions can pursue the plastic deformation while the overall material still remains elastic. A similar effect has been reported by Joseph et al. [27], who have found a localization of strain between Si particles or within Si clusters resulting from the onset of plastic flow at sharp corners of the particles. Although on the macro-scale an averaged effect of the deformation in different phases is observed, on the micro-scale, an input of plastified regions in the Al matrix is much more pronounced. Thus, an onset of the plastic deformation can be detected on the stress-strain curves from FEM simulations at very small strain levels already.

4. Conclusions and outlook

By carrying out the mechanical tests on the macro specimens of Al-Si alloys, it has been observed that the DAS values and Si particles density correlate better with the measured strength characteristics of the alloys than 2D size-shape features of Si particles. These results have made evident the limits of the approach based on 2D characterization of Si morphology and consideration of features of only one phase while assessing the properties of a two-phase material. Hereby, the advantages and limitations of the experimental and numerical methods of mechanical characterization of Al-Si alloys have been disclosed. In particular, the macroscopic mechanical behavior of the alloys has been correlated with the properties of the primary phase (DAS) whereas the numerical simulations have allowed to assess the impact of 3D morphological properties of the eutectic phase. As a result, the combination of both methods of mechanical characterization has enabled to interpret the behavior of the material observed on the macroscopic level.

Thus, the morphology of both phases has been considered simultaneously when analyzing the results of mechanical tests. In this respect, it has been suggested that the density of Si particles and the size of Al dendrites are more important for the strengthening of the alloys than the size-shape features of the eutectic Si induced by the modification. Besides, it has been suggested that the advantages of the modified Si morphology for the ductility of the alloys might be overridden by the properties of the primary phase.

In order to deduce a morphological scenario of an optimal structure of Al-Si alloys for a certain application, further investigations are required: in particular, the investigations of the damage mechanism, consideration of residual stresses in the particles and their failure criteria during FEM simulations as well as analysis of properties and mechanical impact of the primary phase and its interplay with the eutectic phase morphology. Although it has been suggested that both phases exert an influence on mechanical properties, the extent to which they do so is still not elaborated. It might be possible that one of the phases has a dominating mechanical impact under certain conditions or from a certain amount. Besides, one should bear in mind that the results issued from laboratory experiments and computer simulations cannot be compared directly: Compression and tension tests are performed on the two-phase specimens while FEM simulations are usually limited to the eutectic phase only.

References

[1] Kores S, Zak H, Tonn B. Aluminium alloys for cylinder heads. *RMZ – Mater Geoenvironment* 2008; 55: 307–317.

[2] Ye H. An Overview of the Development of Al-Si-Alloy Based Material for Engine Applications. *J Mater Eng Perform* 2003; 12: 288–297.

[3] Hafiz MF, Kobayashi T. A study on the microstructure-fracture behavior relations in Al-Si casting alloys. *Scr Metall Mater* 1994; 30: 475–480.

[4] Requena G, Garcés G, Rodríguez M, et al. 3D architecture and load partition in eutectic Al-Si alloys. *Adv Eng Mater* 2009; 11: 1007–1014.

[5] Requena G, Garcés G, Asghar Z, et al. The Effect of the Connectivity of Rigid Phases on Strength of Al-Si Alloys. *Adv Eng Mater* 2011; 13: 674–684.

[6] Kruglova A, Engstler M, Gaiselmann G, et al. 3D connectivity of eutectic Si as a key property defining strength of Al–Si alloys. *Comput Mater Sci* 2016; 120: 99–107.

[7] Caceres CH, Griffiths JR. Damage by the cracking of silicon particles in an Al-7Si-0.4Mg casting alloy. *Acta Mater* 1996; 44: 25–33.

[8] Wang QG, Cáceres CH. The fracture mode in Al–Si–Mg casting alloys. *Mater Sci Eng A* 1998; 241: 72–82.

[9] Wang QG. Microstructural Effects on the Tensile and Fracture Behavior of Aluminum Casting Alloys A356 / 357. *Metall Mater Trans A* 2003; 34: 2887–2899.
[10] Cáceres CH, Davidson CJ, Griffiths JR. The deformation and fracture behaviour of an Al-Si-Mg casting alloy. *Mater Sci Eng A* 1995; 197: 171–179.
[11] Bai Y, Zhao H. Tensile properties and fracture behavior of partial squeeze added slow shot die-cast A356 aluminum alloy. *Mater Des* 2010; 31: 4237–4243.
[12] Hafiz MF, Kobayashi T. Fracture toughness of eutectic Al-Si casting alloy with different microstructural features. *J Mater Sci* 1996; 31: 6195–6200.
[13] Doglione R. In situ investigations on the ductility of an Al-Si-Mg casting alloy. *JOM* 2012; 64: 51–57.
[14] Cáceres CH, Svensson IL, Taylor JA. Strength-Ductility Behaviour of Al-Si-Cu-Mg Casting Alloys in T6 Temper. *International Journal Of Cast Metals Research* 2003; 15: 531–543.
[15] Russ JC, Dehoff RT. *Practical Stereology*. Boston, MA: Springer US, 2000. Epub ahead of print 2000. DOI: 10.1007/978-1-4615-1233-2.
[16] Roland M, Kruglova A, Harste N, et al. Numerical Simulation of Al-Si Alloys With and Without a Directional Solidification. *Image Anal Stereol* 2014; 33: 29–37.
[17] *ASM Handbook Volume 15: Casting*. ASM International, 1988.
[18] CEN Europäisches Komitee für Normung. *Kunststoffe - Bestimmung der Zugeigenschaften - Teil 2: Prüfbedingungen für Form- und Extrusionsmassen (EN ISO 527-2:1993 einschließlich Corr 1:1994)*. 1996.
[19] Fat-Halla N. Structural modification of Al-Si eutectic alloy by Sr and its effect on tensile and fracture characteristics. *J Mater Sci* 1989; 24: 2488–2492.
[20] Vander Voort GF, Asensio-Lozano J. The Al-Si Phase Diagram. *Microscopy and Microanalysis* 2009; 15: 60.
[21] Fatahalla N, Hafiz M, Abdulkhalek M. Effect of microstructure on the mechanical properties and fracture of commercial hypoeutectic Al-Si alloy modified with Na, Sb and Sr. *J Mater Sci* 1999; 34: 3555–3564.
[22] Liao H, Sun Y, Sun G. Correlation between mechanical properties and amount of dendritic α-Al phase in as-cast near-eutectic Al-11.6% Si alloys modified with strontium. *Mater Sci Eng A* 2002; 335: 62–66.
[23] Gruzleski JE, Closset BM. *The Treatment of Liquid Aluminum-Silicon Alloys*. Des Plaines, Illinois: American Foundrymen's Society, Inc., 1990.
[24] Shivkumar S, Ricci S, Keller C, et al. Effect of solution treatment parameters on tensile properties of cast aluminum alloys. *J Heat Treat* 1990; 8: 63–70.
[25] Poole WJ, Dowdle EJ. Experimental measurements of damage evolution in Al-Si eutectic alloys. *Scr Mater* 1998; 39: 1281–1287.
[26] Guiglionda G, Poole WJ. The role of damage on the deformation and fracture of Al-Si eutectic alloys. *Mater Sci Eng A* 2002; 336: 159–169.
[27] Joseph S, Kumar S, Bhadram VS, et al. Stress states in individual Si particles of a cast Al-Si alloy: Micro-Raman analysis and microstructure based modeling. *J Alloys Compd* 2015; 625: 296–308.
[28] *ASM Handbook Volume 8: Mechanical Testing and Evaluation*. ASM International, 2000.
[29] Closset B, Gruzleski JE. Structure and properties of hypoeutectic Al-Si-Mg alloys modified with pure strontium. *Metall Trans A* 1982; 13: 945–951.
[30] Lee FT, Major JF, Samuel FH. Effect of Silicon Particles on the Fatigue Crack Growth Characteristics of A1-12 Wt Pct Si-0 . 35 Wt Pct Mg- (0 to 0 . 02) Wt Pct Sr Casting Alloys. 1995; 26: 1553–1570.
[31] Haque MM. Effects of strontium on the structure and properties of aluminium-silicon alloys. *J Mater Process Tech* 1995; 55: 193–198.
[32] Odgaard A, Gundersen HJG. Quantification of connectivity in cancellous bone, with special emphasis on 3-D reconstructions. *Bone* 1993; 14: 173–182.

IOP Conf. Series: Materials Science and Engineering **258** (2017) 012004 doi:10.1088/1757-899X/258/1/012004

[33] Agarwal H, Gokhale AM, Graham S, et al. Quantitative characterization of three-dimensional damage evolution in a wrought Al-alloy under tension and compression. *Metall Mater Trans A* 2002; 33: 2599–2606.
[34] *ASM Handbook Volume 19: Fatigue and Fracture*. ASM International, 1996.
[35] Kumai S, Tanaka T, Zhu H, et al. Tear Toughness of Permanent Mold Cast and DC Cast A356 Aluminum Alloys. 2004; 45: 1706–1713.
[36] Roland M, Kruglova A, Gaiselmann G, et al. Numerical simulation and comparison of a real Al-Si alloy with virtually generated alloys. *Arch Appl Mech* 2015; 85: 1161–1171.
[37] Joseph S, Kumar S. Finite Element Analysis of Stress Evolution in Al-Si Alloy. *J Mater Eng Perform* 2015; 24: 253–260.

IOP Conf. Series: Materials Science and Engineering **258** (2017) 012005 doi:10.1088/1757-899X/258/1/012005

Hot cracking of Structural Steel during Laser Welding

Rosa M. Pineda Huitron[1,a*], **Esa Vuorinen**[1,b]

Luleå University of Technology, Division of Materials Science, 97187, Luleå, Sweden

[a*]rosa.maria.pineda.huitron@ltu.se, [b]esa.vuorinen@ltu.se

Abstract. Laser welding is an important technique in many industries due to its high precision in operation, its local and fast processing, narrow welds and its good weld surface quality. However, the process can involve some complications due to the rapid heating and cooling of the material processed, resulting in physical and metallurgical effects as thermal contraction during solidification, giving as a result the presence of residual stresses in the narrow weld. Formation of defects during the process is an important topic to be evaluated in order to achieve better performance of the steels in use. In the present work, defects formed during laser welding of a structural steel have been investigated. The defects formed have been identified and the causes of the defects are discussed. Possible strategies for improvement of the welding procedure and final weld result are proposed. The defects were analysed by optical and scanning electron microscopy and hardness measurement. Cracks were located in the middle of the fusion zone and followed both inter-granular and trans-granular paths. Impurities as manganese sulphides were found along the welding direction, and could act as sites for crack formation. The cracks formed during solidification of the weld are identified as solidification cracks. This kind of cracks is usually caused by solidification shrinkage and thermal contractions during the process, which appear in the fusion zone and sometimes in the heat affected zone.

1. Introduction

The laser welding process is considered important in different industrial applications due to its ability to achieve narrow welds and good weld surface quality. The process has high precision and it is a fast processing method. For many industries, control of the laser welding process has become a challenge in order to be able to produce uniform weld quality without affecting the mechanical properties of the material [1].The demands to create high strength and high quality steels for automotive applications have increased by the years. With regard to the increased demands, defects formed during laser welding process are a very important topic to understand and control in order to produce welds with high strength and quality.

1.1. The laser welding process

The laser welding represents a delicate balance between heating and cooling within a spatially localized volume overlapping two or more solids, in such way that a liquid pool is formed and remains stable until solidified. The main goal of applying laser welding is to create the liquid melt pool by absorption of incident radiation, allow it to grow to the desired size, and then to propagate this melt pool through the solid interface in order to join the two components together. Unsuccessful result is obtained if the melt pool is too large or too small or if significant vaporization occurs while it is present. The quality of the weld can be deteriorated by vaporization of alloy components, excessive thermal gradients that lead to cracking during solidification and instabilities in the volume and geometry of the weld pool that can result in porosity and void formation [2].

1.2. Crack formation mechanisms

Crack formation during the welding process can appear due to the rapid heating and cooling of the material processed, resulting in physical and metallurgical effects. When the material solidifies in the fusion zone, different complex phenomena occur and, both the fusion zone and the surrounding areas can be affected by thermal expansion and contraction, resulting in residual stresses in the narrow weld [3].

1.3. Metallurgical aspects

Different metallurgical factors can affect the quality of laser weld. These can be beneficial or detrimental for crack formation. Beside the internal stresses, can solidification modes, phase transformations, micro- and macro-segregation of elements during solidification and kinetics of precipitates caused by micro-alloying elements influence the weld quality [4-5].The effect of alloying elements is another important topic in the evaluation of causes for crack formation during laser welding process. Impurities cause severe cracking in carbon and low-alloy steels even at relative low concentrations. The focus of this work has been to investigate the causes of crack- and defect-formation during laser welding of two structural steels, used in automotive applications. In the work has different laboratory characterization techniques been used and, combined with simulations by a specific software. The work has in addition to identification of defects and their causes, included identification of strategies for improvement of the laser weld quality [6-8].

2. Materials and Methods

2.1. Materials

Laser welded structural steels, mainly used for automotive components, were supplied by a Scandinavian company. The chemical composition of the two steels welded by laser is indicated in Table 1. The steels were laser welded against each other and, the resultant specimen was sectioned from the middle part, in the fusion zone (FZ). The specimen was analyzed longitudinally and perpendicularly to welding direction. This work was mainly focused on the identification of and causes for the defects formed during or after the laser welding process. Samples were prepared for metallographic analysis following the standard sample preparation process by grinding and polishing. The final polishing step was done by using colloidal silica suspension. In addition, the samples were etched with Nital 3%. In addition, a sample before welding was taken as a reference in order to compare the surface and the microstructure.

Table 1. Chemical composition of the laser welded structural steels

%	C	Si	Mn	P	S	Nb	V	Al	Ti
Steel 1	0.139	0.190	1.550	0.018	0.004	0.037	0.097	0.042	0.022
Steel 2	0.29	0.25	0.75	0.035	0.035	-	-	-	-

2.2. Methods

2.2.1. Characterization techniques. Microstructure analysis of the specimens was carried out by using Nikon MA200 Optical Microscopy and Scanning Electron Microscopy (SEM JEOL IT 300). In the same way, SEM was used for applying the Energy Dispersive X-Ray Spectroscopy (EDX) for the qualitative and quantitative elemental analysis.

IOP Conf. Series: Materials Science and Engineering **258** (2017) 012005 doi:10.1088/1757-899X/258/1/012005

2.2.2. X-Ray Diffractometry (XRD). In XRD observations were monochromatic Cu-Kα radiation with 40 kV and 45 mA used. The anode diffractometer was used to scan the angular 2θ range of 20-120 degrees with scanning time of 20 minutes per each specimen. The resulting profiles were analysed by the High score plus software to obtain the exact position of the peaks. Peak positions and intensities of (111), (112), (200), (211), (220) and (130) planes were identified. Together with the Bragg's law equation and the lattice parameters obtained, it was possible to calculate 2θ values of austenite, ferrite and martensite [10].

2.2.3. Microhardness test by Vickers. Measurements were performed in the welding direction across the FZ following the area close to the defects. Measurements in the cross section follows the FZ, the heat affected zone (HAZ) and the base metal (BM). The load applied was 100 g load ($HV_{0.1}$).

2.2.4. Thermodynamic calculations. Prediction of phases under equilibrium conditions was calculated by using a commercial software (Thermo-calc). This program gives a general idea of the main phases formed during equilibrium conditions. Therefore, it was possible to calculate the presence of phases formed during solidification in local equilibrium condition with help of Scheil curves. In this simulation, some assumptions were made; no diffusion in the solid phase and fast diffusion in the liquid and it is assumed homogeneous. Calculation of the temperature-time-precipitation (TTP) curves was performed by TC-Prisma integrated in Thermo-calc. For this calculation, homogeneous diffusion of elements in the FZ was assumed. The purpose of this calculation was to present and compare, with the experimental analysis, the formation of the phases in order to find a way to relate the results with the behavior of the material during the laser welding process.

3. Results and discussions

3.1. Microstructural analysis

The microstructure of this material before welding consists of ferrite (F) and pearlite (P) in a banded structure, Fig. 1a. The microstructure after laser welding process consists of a mixture of tempered martensite (TM) and ferrite. Presence of untempered martensite (UM) was located mainly in FZ. Additionally, due to bainite (B) is difficult to distinguish from TM, there is a possibility to find some areas with B present in grain boundaries. Fig. 1b. The presence of UM is related to the rapid heating and cooling of the steel during the process. However, UM located in the FZ can be one of the reasons for crack formation during the laser welding process.

3.2. Identification of defects

Initially, defects found in the bulk, before laser welding process were located in the middle of the banded hot rolled structure, which correspond to manganese sulfide inclusions, normally present in this kind of steels. Fig. 2. The size of the defects was between 50-100 μm. Nevertheless, slags from the casting process can form some inclusions due to the short times at a high temperature or longer time at a moderated temperature during the process. In addition, cracks are located in the FZ on the steel after laser welding following trans- and inter- granular directions. Fig. 3. The length of the cracks are between 2.5 – 3.9 mm with 1 mm depth. Inclusions were found in the middle of the crack. Fig 3(b). These inclusions are identified as oxides because they are non-conductive. The analysis of the crack was performed in the cross section, perpendicularly to the welding direction. The FZ, HAZ and BM were identified, Fig. 4. It is possible to observe a difference shape of grains from the FZ to the BM. In all cases, the cracks follow the grain boundaries but some cracks are crossing the grains following random directions, inter- and trans-granular cracks respectively.

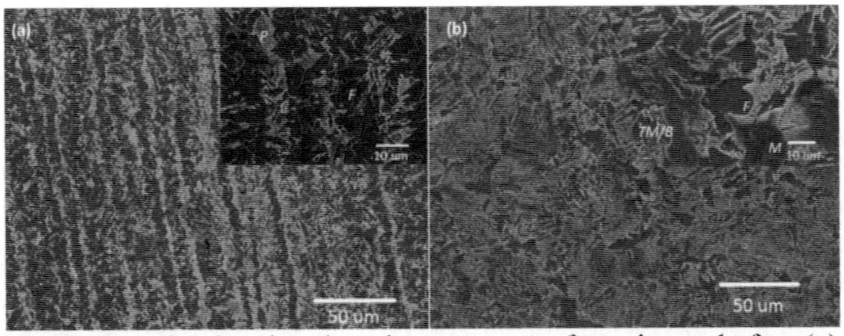

Figure 1. SEM micrographs showing the microstructure of specimens before (a) and after (b) laser welding process.

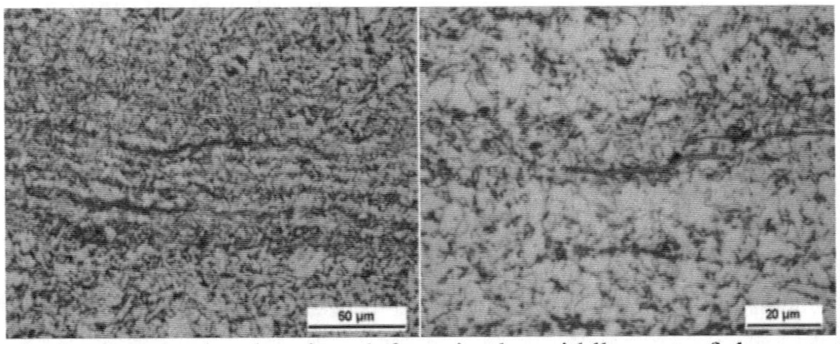

Figure 2. Optical micrographs showing defects in the middle part of the sample following the hot rolled marks before laser welding process.

Cracks presented above were created during the process defined as solidification cracking mechanism. This kind of defects appear when the steel solidified caused by both shrinkage - and thermal-contraction. This kind of cracks normally form along the FZ grain boundaries but sometimes in the sub-grains. Also, is normally caused by the presence of stresses during the process; not to mention the effect of alloying elements and impurities on the weld which also affect the behavior of the steel during welding creating cracks. For this particular case, the location and direction of the cracks found is special and, those are not usually found in this kind of steels due to their low susceptibility to cracking in laser welding. However, the information and, the number of articles treating solidification cracking on structural steels during laser welding is limited [1-4]. Different causes can be detrimental for cracking formation in this kind of steels as the presence of brittle micro-constituent, in this case untempered martensite (UM) which was found in the FZ, and impurities formed during previous processing steps, which lead to segregation of MnS and inclusions. The kinetics of precipitation formation is very difficult to control because the fast heating and cooling during laser welding process. Thus, there is not enough time for precipitates to be formed by diffusion during the solidification and, there is not either time for existing precipitates to be dissolved during the heating cycle, and these can act as sites for crack formation. Finally, the control of process parameters is very important in order to have better control of the kinetics during solidification process such as heat input, laser power and welding speed [7-8].

3.3. Vickers microhardness tests

Vickers micro-hardness measurements using load of 100 g were performed. Measurements were applied to the samples before- and after- laser welding process in the longitudinal- and in the cross-sections with regard to the welding direction, respectively, Fig. 5 a. The distribution of the hardness values was uniform before welding, having an average of 214 HV (from 197 to 257 HV). The maximum values correspond to pearlite, while the lower values correspond to ferrite micro-constituents, Fig. 5c. For the welded samples, higher hardness values were found (from 218 to 352 HV). Presence of UM and TM might explain the highest hardness values. The FZ exhibit up to 352 HV, this value correspond to the presence of UM, while the HAZ exhibit up to 300 HV corresponding to TM, Fig. 5b. Presence of UM is known for having an important influence on crack formation during laser welding.

3.4. Chemical and Phase analysis

Qualitative and quantitative chemical analysis by SEM-EDX was performed. In addition, XRD technique was applied to control possible phase changes during the laser welding process. EDX analysis exhibited the concentration of MnS, just for the specimen before welding with low presence of alumina, Fig. 6a. For the samples after laser welding, SiO_2 was found inside the cracks located in the FZ for both, longitudinal- and cross-section to the welding direction. These inclusions might affect the behavior of the material during solidification and act as sites for concentration of stresses and crack formation. The integrated peaks position, from XRD results, correspond to the planes (111), (112), (200), (211), (220), (130) for the three specimens before welding and FZ and cross section after welding as shown in Fig. 7a. The phases present in the material correspond mainly to retained austenite (111), (020) and (022) peaks and, to ferrite (110), (200), (211), (220) and (103) peaks, Fig. b-c. The identification of martensite with XRD technique is complex because the 2θ values are very close to those that correspond to ferrite. For the identification of martensite, in this case, some calculations were done by using the plane distance equation for tetragonal crystal cell, the lattice parameter and the Bragg's equation. The presence of martensite could be related to (200) plane for the laser welded sample in the FZ, because this peak is showing wider, Fig. 7b. However, with this method, it was not possible to identify the presence of MnS but comparison with literature gave an idea of the possible location of these precipitates [10-11].

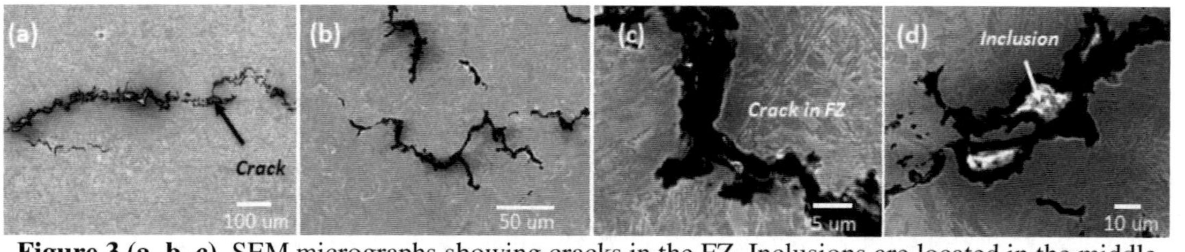

Figure 3 (a, b, c). SEM micrographs showing cracks in the FZ. Inclusions are located in the middle of the cracks (3c).

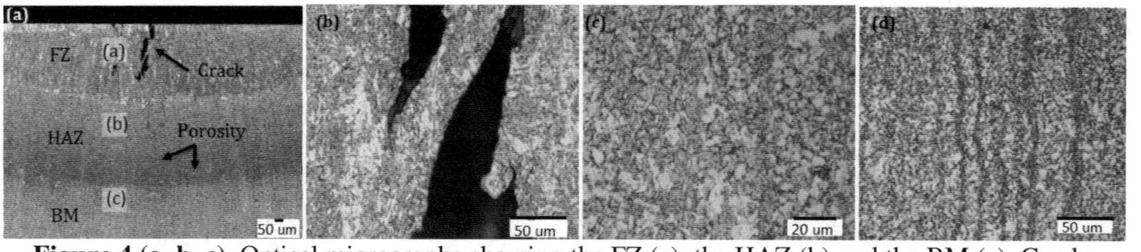

Figure 4 (a, b, c). Optical micrographs showing the FZ (a), the HAZ (b) and the BM (c). Cracks located in the FZ (4c).

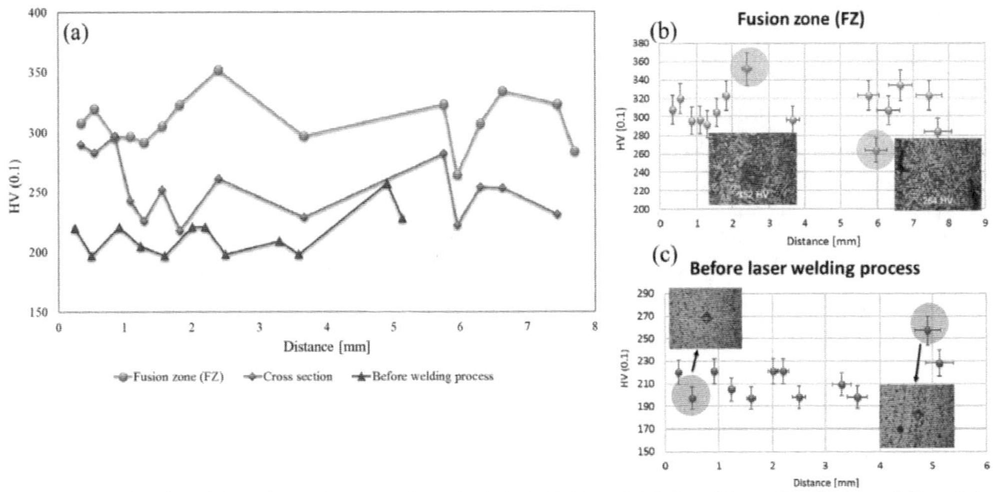

Figure 5. Distribution of Hardness values in FZ, HAZ and BM (a-b) and before welding (c).

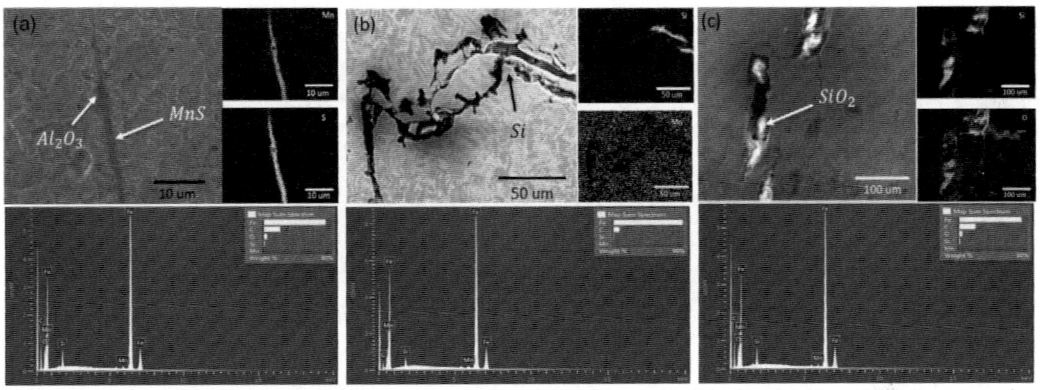

Figure 6. Chemical analysis by EDX on specimens before (a) and after laser welding process in the (b) longitudinal direction (FZ) and the (c) cross section.

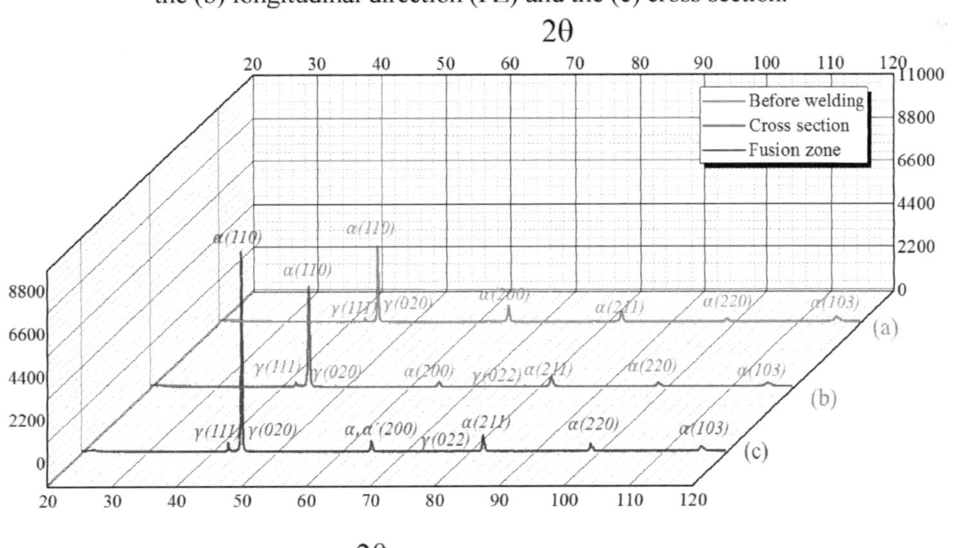

Figure 7. XRD analysis data showing the (a) comparison of the specimens (b) before and (c) after laser welding process.

3.5. Thermodynamic calculations

Phases in equilibrium were calculated by setting the initial chemical composition and the proper thermodynamic database in thermo-calc software. Phases as austenite, ferrite, and cementite could be observed, Fig. 8. Further, precipitates of elements such as V, Nb, Mo, Ti and Mn can appear for this material. Those precipitates can be present in the material as carbides, nitrides or carbo-nitrides. MnS was another interesting phase. According of the experimental results, segregation of manganese sulfides were found in this steel (in hot rolling condition. Moreover, the segregation of this phase was observed to appear in the range of temperature from 1100 to 500°C, Fig. 9. In addition, precipitation kinetics was simulated by TC-Prisma by using assumptions as nucleation sites in grain boundaries, having FCC as matrix and interfacial energy of 0.5 J/m. Fig. 10 shows the formation of precipitates such as V and Nb (C, N) and MnS. The results give us a wide view of the formation of precipitates for this structural steel. Precipitation of V (C, N) is faster in comparison with Nb (C,N) and MnS precipitates, this forms in the range of 850-970 °C in approximately 10 seconds. However, a more detailed thermodynamic study of the precipitation and segregation formation for this material and its effect on crack formation after laser welding is recommended. Similarly, Scheil calculation was performed to simulate the evolution of phases during solidification process and to estimate the solute redistribution between the solid and the solidifying liquid phase. The solid fraction of delta phase appear at higher temperatures, that is related with the peritectic transformation because the amount of carbon in this steel. The solid fraction of austenite change from higher to lower temperatures while some other phases appear [13]. The segregation of MnS appears between 1350–1100 °C. Other elements as titanium carbo-sulfide ($Ti_4C_2S_2$) and V (C, N) was observed. Fig. 11 [12-13].

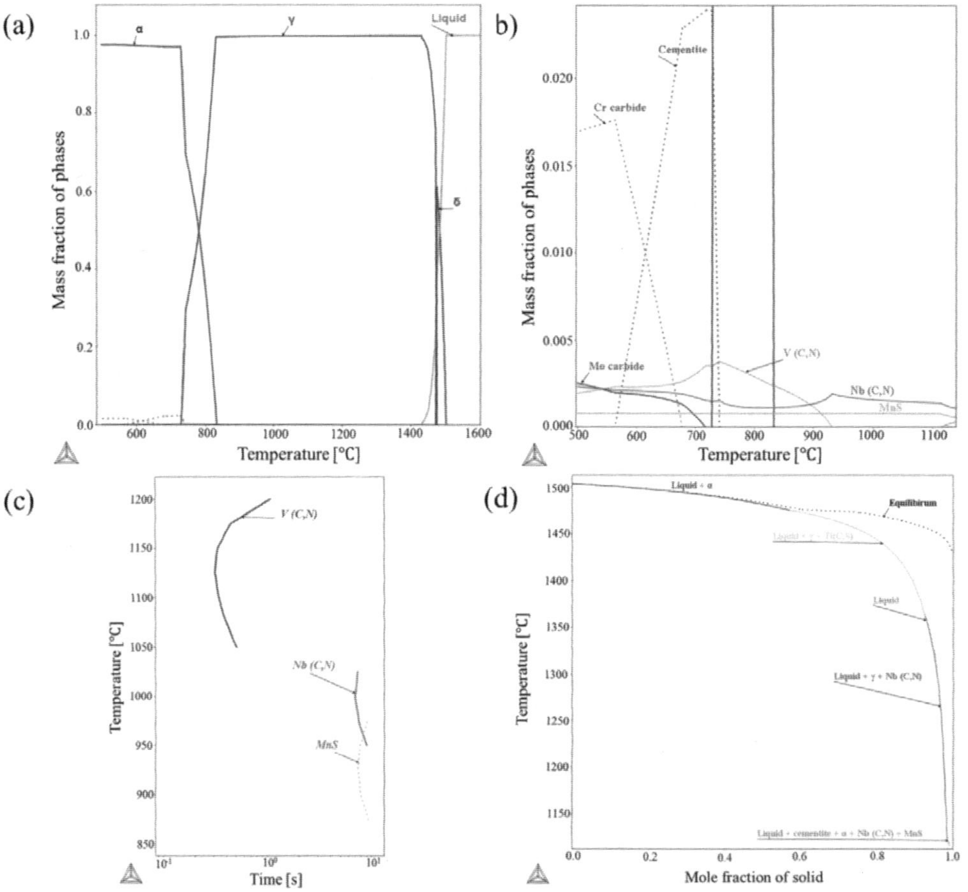

Figure 8. Thermodynamic calculations of the (a) formation of phases in equilibrium, (b) Precipitates in equilibrium, (c) temperature-time-precipitation and (d) the scheil curve.

4. Conclusions

A complete study was carried out in order to identify the defects and their causes during laser welding of structural steels. The microstructural analysis of the steel before and after laser welding process was successfully performed. The hardness of the microstructures has been measured and related with the phases formed before and after laser welding. The Chemical- and phase- analysis by using SEM-EDX and XRD techniques was included. Thermodynamic calculations were done for predicting and comparing the phases formed in this steel in equilibrium and non-equilibrium conditions by Thermo-calc software. This in order to explain the performance of the steel during laser welding process.

Cracks were formed in the center of the fusion zone, longitudinally oriented, which is unusual orientation, following inter- and trans- granular directions. Those cracks were identified as solidification cracks. Causes of solidification cracks were related to the presence of untempered martensite formed in the fusion zone, which increases the risk for crack formation during solidification. The high concentration of Mn and S found in this steel before welding increases also the susceptibility for crack formation. Precipitates containing Si and O have been found on the crack surfaces and, their origin has not been possible to explain in this work.

The presence of these elements together with the brittle micro-constituent and the fast solidification during laser welding process, create tensile stresses by volume shrinkage in the fusion zone, causing the formation of the cracks identified.

Acknowledgments

Authors want to thank to the National Council on Science and Technology of Mexico (CONACYT) for the received support.

References

[1] Duley W W 1999 *Laser Welding* (Canada, Wiley-Interscience publication) pp 3-66
[2] Kou S 2003 *Welding Metallurgy* (Wiley-Interscience publication) pp 263-320
[3] Lippold J C 2015 *Welding Metallurgy and Weldability* (New Jersey: Wiley) pp 84-129
[4] Böllinghaus T, Herold H 2011 *Hot cracking fenomena in welds III* (Berlin: cross editors Springer) pp 145-161
[5] Hernandez S 2010 *Laser Welding of Hardenable steel* Master thesis (Luleå Sweden: Luleå University of Technology)
[6] Liu J and Kou S 2015 Effect of diffusion on susceptibility to cracking during solidification *Acta Materialia* **100** pp 359-368
[7] Phaoniam R, Shinozaki K, Yamamoto M, Kadoi K, Nishijima A and Yamamoto M 2014 Solidification cracking susceptibility of modified 9Cr1Mo steel weld metal during hot-wire laser welding with a narrow gap groove *Welding in the World* **58** pp 469-476
[8] Yilbas B and Akhtar S 2013 Laser welding of AISI 316 steel: Microstructural and stress analysis *Journal of Manufacturing Science and Engineering* **135**
[9] *Metallography and Microstructures ASM HAndbook* 2004 (**9** ASM International)
[10] Cullity B D and Stock S R *Elements of X-Ray Diffraction* (Third Edition) pp 347-355
[11] Pietryka I and Lisak J 2010 The identification of carbide phases by XRD analysis as the method of assess the extend of the steel damage after long time in service AFE **10** pp283-288
[12] Lukas H L, Fries S G and Sundman B 2007 *Computational Thermodynamics The Calphad Method* pp 243-262
[13] Guimaraes M, Vasconcellos da Costa e Silva A L 2015 Evaluation segregation in HSLA steels using computational thermodynamics *J. Mat. Re.s and Tech.* **4** pp 353-358

IOP Conf. Series: Materials Science and Engineering **258** (2017) 012006 doi:10.1088/1757-899X/258/1/012006

Frictional Performance Assessment of Cemented Carbide Surfaces Textured by Laser

S Fang[1,2,3], L Llanes[1,2,*], S Klein[3], C Gachot[4], A Rosenkranz[4], D Bähre[3], F Mücklich[4]

[1]CIEFMA – Dept. Materials Science and Engineering, Universitat Politècnica de Catalunya, EEBE-Campus Diagonal Besòs, 08019 Barcelona, Spain
[2]Barcelona Research Center in Multiscale Science and Engineering, Universitat Politècnica de Catalunya, 08019 Barcelona, Spain
[3]Institute of Production Engineering, Saarland University, Saarbrücken, 66123 Germany
[4]Chair of Functional Materials, Saarland University, Saarbrücken, 66123, Germany

* luis.miguel.llanes@upc.edu

Abstract. Cemented carbides are advanced engineering materials often used in industry for manufacturing cutting tools or supporting parts in tribological system. In order to improve service life, special attention has been paid to change surface conditions by means of different methods, since surface modification can be beneficial to reduce the friction between the contact surfaces as well as to avoid unintended damage. Laser surface texturing is one of the newly developed surface modification methods. It has been successfully introduced to fabricate some basic patterns on cemented carbide surfaces. In this work, Direct Laser Interference Patterning Technique (DLIP) is implemented to produce special line-like patterns on a cobalt (Co) and nickel (Ni) based cemented tungsten carbide grade. It is proven that the laser-produced patterns have high geometrical precision and quality stability. Furthermore, tribology testing using a nano-tribometer unit shows that friction is reduced by the line-like patterns, as compared to the polished one, under both lubricated and dry testing regimes, and the reduction is more pronounced in the latter case.

Key words: Cemented carbide; Laser Surface Texturing (LST); Direct Laser Interference Patterning Technique (DLIP); Tribology; Friction.

1. Introduction

Cemented tungsten carbide hardmetal is made of tungsten carbide powder (WC) with metallic powder by sintering under high pressure and temperature. Due to its high hardness and suitable toughness, hardmetals are often used as cutting and forming tool materials (e.g. Ref. [1]). Electrical Discharge Machining (EDM) is one of the most advanced hardmetal processing technologies applied in the industry. EDM is capable to shape the material with relatively precise dimensions, but some thermal defects can be induced such as pores and re-disposition [2-5]. Therefore, it is difficult to obtain the desired surface structure with higher accuracy using this method. Laser surface texturing (LST) has a wide range of application as an approach of surface modification. It can be an alternative technique to ablate the hardmetals and to modify their surface structure with accuracy in micrometer (μm) scale.

In the LST family, Direct Laser Interference Patterning Technique (DLIP) is a metallurgical process that uses a superposition of two or three pulsed laser beams with the same frequency [6,7]. The laser beam ablation is mainly dependent on three factors: wavelength, energy density and duration of pulses. The wavelength of the laser beams has crucial impact on the energy absorption. In general, for metals and ceramics the absorption is higher with shorter waves. The absorption in the case of the cobalt based cemented tungsten carbide is for example 85% at 355 nm [8]. The ablation rate of the laser beams is related to the energy density and the pulse duration decides the mechanism of laser-matter reaction [7]. The incident laser beam generates an electric field, which excites the electrons, and the energy is then transmitted to the electrons by acceleration. The electrons collapse the lattice and transfer the energy to the atoms. The vibration of lattice causes the material heating and phase transformation. The energy transfer needs about 1ps for metals and slightly more for ceramics [9]. Therefore, for the ns-laser, the energy transfer takes place mostly through melting and vaporization [10]. Due to the high intensity, energy transformation of the laser beams to the material surface, i.e. melting, recrystallization, phase transformation etc., occurs in the scale of micro-/nanometer. Therefore, the surfaces generated by laser always exhibit clean and precise.

It is indicated that some basic patterns are produced on the hardmetal surface such as dimples and grooves [8,11,12]. However, there is not enough research work available dealing with the surface modification of hardmetals for tribological applications. Therefore, the work described the production of specific line-like patterns on cemented carbide by means of DLIP and the influence of the patterns on frictional performances is assessed on a nano-tribometer.

2. Materials & Experimental Procedure

2.1. Materials

A Cobalt Cemented Tungsten Carbide grade, in which the embedded WC grains have an average size of 20 μm and the binder (Co+Ni) has a nominal amount of 28 wt. %, has been selected as test samples. The microstructural characteristics of the selected material are displayed in Table 1. All the test samples should be fine polished prior to the laser treatment. After the polishing, the microstructure of the hardmetal is exposed. It is observed that the grains are randomly distributed and some damage is caused by polishing such as cracks or holes, but they can be ignored (Figure 1).

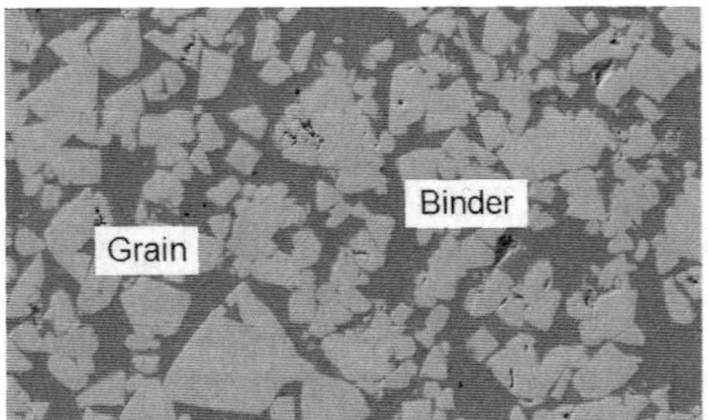

Figure 1. Polished WC-CoNi hardmetal surface, characterized by scanning electron microscopy (SEM)

IOP Conf. Series: Materials Science and Engineering **258** (2017) 012006 doi:10.1088/1757-899X/258/1/012006

Table 1. Microstructural characteristics and properties of the used WC-CoNi hardmetal.

WC-Grain Size (μm)	Co (wt. %)	Ni (wt. %)	Density (g/cm³)	Hardness (HV30)
20	14	14	12.82	610

2.2. Experimental Parameters

A special laser trajectory is constructed to create the interference patterns (Figure 2(a)). The laser beam is split into two sub-beams, which are conducted by a series of mirrors and converged at the focus point, where the beams produce the interference phenomena. The line-like structures are produced when these phenomena occurs on the material surface. The information of the laser installation is shown in Table 2.

Table 2. Information and configuration of the ns-laser installation.

Laser type	Laser source	Pulse duration	Wave length (nm)	Pulse repetition frequency (Hz)	Fluence (J/cm²)	θ(°)
ns-laser	Nd: YAG	10 ns	355	10	2.3	1.8

The periodicity of the interference patterns can be calculated with Equation (1) [6,7]:

$$P = \frac{\lambda}{2sin(\frac{\theta}{2})} \quad (1)$$

λ: Laser wavelength
θ: Angle between two sub-beams
P: Calculated periodicity

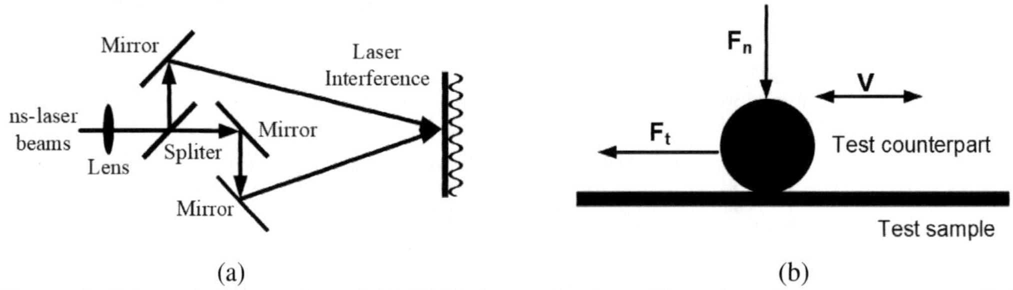

Figure 2. Schematic illustration of (a) DLIP for production of laser interference structure and (b) dynamics of the nano-tribometer.

According to the configuration parameters of the ns-laser installation in Table 2 and Equation (1), the line-like patterns have the periodicity of 11.3 μm.

2.3. Frictional Tests on the Nano-tribometer

The nano-tribometer (NTR-CSM instruments) is used to measure the coefficient of friction (COF) by sliding movement between the sample and balls with or without lubricants. The normal force Fn that is applied on the ball, as well as the tangential force Ft that is produced by the linear reciprocal movement of the sample holder, are recorded during the test (Figure 2(b)). COF is then calculated in account of the two forces using Equation (2):

IOP Conf. Series: Materials Science and Engineering **258** (2017) 012006 doi:10.1088/1757-899X/258/1/012006

$$\mu = \frac{F_t}{F_n} \tag{2}$$

Friction tests have been conducted on two types of surfaces (polished surfaces and line-like patterns) with and without lubricants, for three repetitions. All the samples are tested under identical conditions, as shown in Table 3. The test counterparts are made of Steel 100Cr6 and have the diameter of 3 mm. The sample is fixed on the test table and moves reciprocally at a maximal linear speed of 1 mm/s with an amplitude 0.6 mm. 1000 cycles are set to test the frictions with and without lubricants. The lubricant Kadiol 50 is used, which has a kinematic viscosity υ of 5.6 mm²/s at 40 °C. The normal load applied on test sample is 1 mN.

Table 3. Nano-tribometer configuration.

Acquisition rate (Hz)	Trajectory	½ Amplitude (mm)	Max Lin.Speed (mm/s)	Frequency (Hz)	Cycle	Normal load (mN)
100	Sinus	0.3	1	0.53	1000	1

3. Results and Discussion

3.1. Geometrical Characterization

The geometrical properties of the line-like patterns are measured and characterized by Laser Scanning Microscopy (LSM). Figure 3(a) shows the surface image taken by LSM, and some discontinuity can be observed along the line-like patterns. The discontinuity represents that the excessive melting and vaporization of the material occurred in certain places. This can be resulted from the inhomogeneous distribution of binder and grains, which have different melting and vaporization temperatures. For line-like patterns, three parameters are measured (Figure 3(b)): (1) periodicity, defined as the distance between two adjacent peaks or bottoms; (2) height, defined as the distance between the peak and its adjacent bottom; and (3) peak angle. The textured surface area is about 20 mm x 3 mm. The mean values of periodicity, height and angle, are 11.8 μm, 0.6 μm, and 7.5° respectively. A wave profile of the line-like patterns can be rebuilt by the software Origin 9 with the measurement data obtained by LSM (Figure 3(c)).

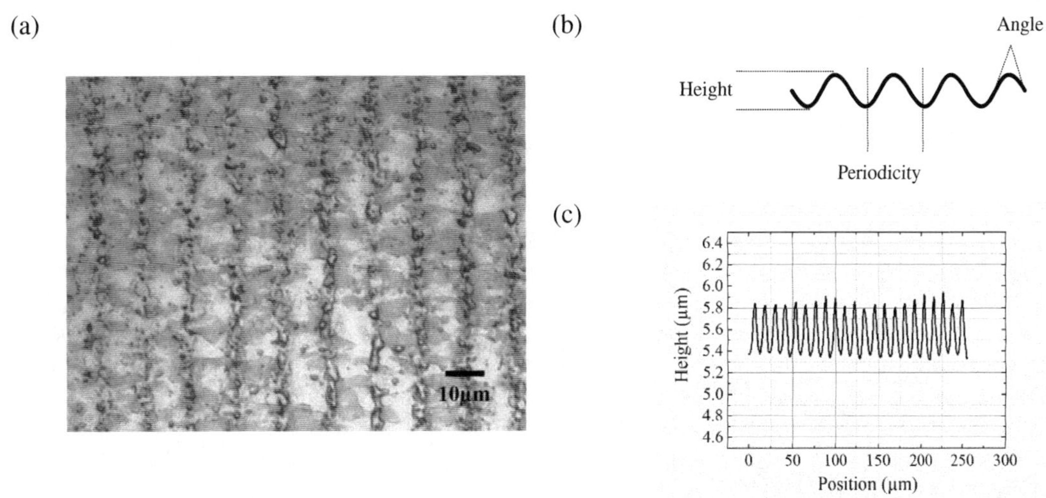

Figure 3. (a) Surface image in 2D (LSM), (b) geometrical characterization of the line-like structure, (c) wave profile of the cross section in the transversal direction.

3.2. Tribological Properties Investigation

The signals of the normal force Fn and the tangential force Ft for each test are recorded, and the real-time values of COFs of the tests are shown in figure 4, where Figure 4(a) and 4(b) show the COF comparison between polished and patterned surfaces in the cases of non-lubricant and lubricated, respectively.

Without lubricants, the frictional performances of the polished and patterned surfaces are quite unstable (Figure 4 (a)). The COF of the polished surface experienced a remarkable rise along with the sliding laps, but the COF of the patterned surface experienced a huge drop at the beginning. These can be explained by the changes of surface microstructural conditions, i.e. deformation of the surface. In the case of polished surfaces, the materials are at first compressed and deformed at the beginning, so that the stacked materials hinder the movement of the samples and cause the COF to increase. However, in the case of patterned surfaces, the surfaces are covered by the line-like structures produced by laser, which are relatively rough than the polished ones; therefore, COF is quite high at the beginning of the test. However, the patterns are deformed and even destroyed along with the sliding laps. After the unstable stage, COF of patterned surface is much lower than the one of the polished surface. The observation can be explained by the fact that the produced debris can work as tiny lubricants between the two surfaces, which may help to reduce friction.

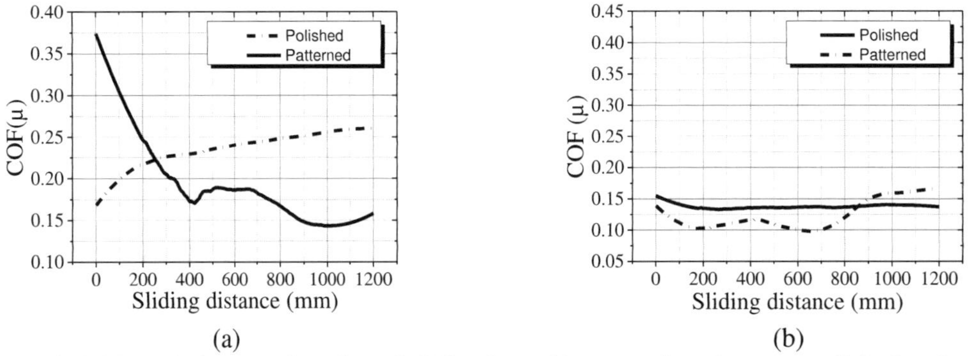

Figure 4. COF evolution as a function of sliding laps: (a) comparison between polished and patterned surfaces in the case of non-lubricant, (b) comparison between polished and patterned surfaces in the case of lubricant.

With lubricants (Figure 4 (b)), the frictional performance of both two types of surface patterns are more stable, except that COF of patterned surfaces slightly increases between 600 cycles and 800 sliding laps. It is supposed that the patterns begin to be destroyed at 600 cycles, and the contact between the two surfaces is becoming less stable till 800 cycles. Then, friction regains stability after 800 cycles, since the patterns are completely demolished and a new stable contact is re-established. The COF of polished surfaces is relatively constant during the whole test except for a slight decrease at the very beginning, so as to the patterned surface. This observation can be explained by the fact that the lubrication film stabilizes the tribological system. However, the difference of the obtained COFs between the patterned and polished surfaces is not very obvious, and it seems that the patterns don't have effective influence in the lubricated case. This observation can be explained by the fact that the discontinuity of the patterns can barely contribute additional hydrodynamic pressure to build up a thick film, which can effectively help reduce the friction in the case of lubricant.

4. Conclusion and Remarks

Despite of the machining difficulty of hardmetals, line-like patterns have been satisfactorily produced on WC-CoNi hardmetal surface by means of LST. It provides a high precision of the structural geometry. The utilization of Direct Laser Interference Patterning Technique (DLIP) allows the creation of the line-like structure on the surface. The structure has a period of 11.8 μm, approximately close to the theoretical prediction of 11.3 μm.

In the frictional tests, it is found that patterns created on the surfaces can effectively reduce the friction compared with the polished surfaces in the non-lubricated case. However, such expectation is not observed in the lubricated case, since the discontinuity of the patterns is unfavourable to supply necessary additional hydrodynamic pressure. Moreover, the patterned surfaces seem to be not as stable as the polished ones in both cases. These observations might be related to the surface integrity and mechanical property change induced by the laser treatment. Therefore, it is advisable to introduce surface integrity assessment, such as micro-indentation, in order to validate the supposition in the future.

Acknowledgements

The work leading to this publication was supported by the German Academic Exchange Service (DAAD) with funds from the German Federal Ministry of Education and Research (BMBF) and the People Program (Marie Curie Actions) of the European Union's Seventh Framework Program (FP7/2007-2013) under REA grant agreement n° 605728 (P.R.I.M.E. – Postdoctoral Researchers International Mobility Experience). The authors are grateful for the material supplier Saar-Hartmetall und Werkzeuge GmbH. Finally, this contribution has also been partly funded by the Spanish Ministry of Economy and Competitiveness through Grant MAT2015-70780-C4-3P (MINECO/FEDER).

References

[1] Prakash L 2014 Fundamentals and General Applications of Hardmetals *Comprehensive Hard Materials* vol 1pp 29–90

[2] Heuvelman.C.J. 1980 Summary report on the CIRP cooperative research on spark-erosion machining of cemented carbides (die-sinking) *CIRP Ann. - Manuf. Technol.* **29** 841–511

[3] Gadalla A M and Tsai W 1989 Machining of WC-Co composites *Mater. Manuf. Process.* **4** 411–23

[4] Llanes L, Casas B, Idanez E, Marsal M and Anglada M 2004 Surface integrity effects on the fracture resistance of electrical-discharge-machined WC-Co cemented carbides *J. Am. Ceram. Soc.* **87** 1687–93

[5] Jahan M P, Rahman M and Wong Y S 2011 A review on the conventional and micro-electrodischarge machining of tungsten carbide *Int. J. Mach. Tools Manuf.* **51** 837–58

[6] Mücklich F, Lasagni A and Daniel C 2006 Laser interference metallurgy - Using interference as a tool for micro/nano structuring *Int. J. Mater. Res.* **97** 1337–44

[7] Duarte M, Lasagni A, Giovanelli R, Narciso J, Louis E and Mücklich F 2008 Increasing lubricant film lifetime by grooving periodical patterns using laser interference metallurgy *Adv. Eng. Mater.* **10** 554–8#

[8] Dumitru G, Lüscher B, Krack M, Bruneau S, Hermann J and Gerbig Y 2005 Laser processing of hardmetals: Physical basics and applications *Int. J. Refract. Met. Hard Mater.* **23** 278–86

[9] Chichkov B N, Momma C, Nolte S, von Alvensleben F and Tünnermann A 1996 Femtosecond, picosecond and nanosecond laser ablation of solids *Appl. Phys. A Mater. Sci. Process.* **63** 109–15

[10] Yao Y L, Chen H and Zhang W 2005 Time scale effects in laser material removal: a review *Int. J. Adv. Manuf. Technol.* **26** 598–608

[11] Dumitru G, Romano V, Weber H P, Sentis M and Marine W 2003 Ablation of carbide materials with femtosecond pulses *Appl. Surf. Sci.* **205** 80–5

[12] Li T, Lou Q, Dong J, Wei Y and Liu J 2001 Modified surface morphology in surface ablation of cobalt-cemented tungsten carbide with pulsed UV laser radiation *Appl. Surf. Sci.* **172** 331–44

Microstructure characterisation of Ti-6Al-4V from different additive manufacturing processes

M Neikter[1*], P Åkerfeldt[1], R Pederson[2] and M-L Antti[1]

[1] Division of Materials Science, Luleå University of Technology, Luleå 971 87, Sweden
[2] Division of Welding Technology, University West, Trollhättan 461 32, Sweden

* magnus.neikter@ltu.se, phone: +46 (0)920-492378

Abstract. The focus of this work has been microstructure characterisation of Ti-6Al-4V manufactured by five different additive manufacturing (AM) processes. The microstructure features being characterised are the prior β size, grain boundary α and α lath thickness. It was found that material manufactured with powder bed fusion processes has smaller prior β grains than the material from directed energy deposition processes. The AM processes with fast cooling rate render in thinner α laths and also thinner, and in some cases discontinuous, grain boundary α. Furthermore, it has been observed that material manufactured with the directed energy deposition processes has parallel bands, except for one condition when the parameters were changed, while the powder bed fusion processes do not have any parallel bands.

1. Introduction

Additive manufacturing (AM) has emerged as a promising manufacturing process for components with complex geometries and low lot sizes. Many different AM processes for metals have been developed [1-7], having different characteristics. Directed energy deposition processes where wire is used are preferred when higher deposition rates are wanted while powder bed fusion processes are better for manufacturing of parts with higher complexity. Titanium has high affinity to oxygen and if not protected in inert atmosphere during manufacturing, this will create a brittle surface layer beneath the metal surface called alpha case [8], which may significantly reduce important mechanical properties. The high oxygen affinity makes the production process of titanium sponge complex, which is one of the reasons why titanium metal is more expensive than e.g. steel. Furthermore, when taking into consideration manufacturing of components, titanium parts become even more expensive in the sense that up to 80% [9] of the original titanium material is machined away when using conventional subtractive manufacturing processes. Additive manufacturing is therefore extra attractive when considering making titanium alloy components. Additive manufacturing takes place in inert atmospheres, usually in argon gas or in vacuum. The loss of expensive titanium metal during AM is low for all processes. For powder bed fusion processes the powder that is not melted during the build can be recycled over and over again, making the final usage up to almost 100%, compared to conventional subtractive manufacturing with buy-to-fly ratios as low as 20% [9]. The typical microstructure of additive manufactured titanium consists of columnar prior beta (β) grains (see section 3.1 for examples of prior β grains) that grow epitaxially [4,10-11] across several layers. When cooling below the β transus temperature alpha (α) phase starts to form, first at the β grain boundaries and then, depending on cooling rate, the α phase grows along the β grain boundaries or into the β grains in a platelet form (see figure 1 for examples of grain boundary α and α laths). The extent of alpha phase at the prior β grain boundaries as well as the size and distribution of α laths inside the prior β grains depend on cooling rate [2,6,12-14]. Faster cooling rates render in less or no alpha phase at the prior β grain boundaries, whereas the size (length/thickness) of individual α laths decreases with increasing cooling rate. Parallel band morphology is sometimes found in AM materials and could easily be interpreted as the deposited layers, which is not the case. As discussed by Kelly et al. [15] the parallel bands are rather a macroscopic phenomenon that is dependent on peak temperature, time at peak temperature and cooling rate. The aim of this work is to present typical microstructures and quantitative results of selected microstructure features that are characteristic for Ti-6Al-4V material built with the five chosen AM processes. It should be pointed out that significant variation of microstructures can be achieved within each AM process by varying the process parameters. As an example of that, Ti-6Al-4V material was built with Laser Metal wire Deposition (LMwD) using two different process parameter settings (LMwD-0 and LMwD-2).

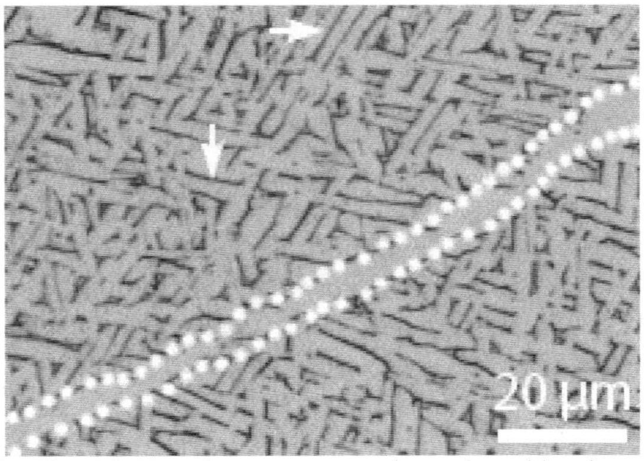

Figure 1. An example of a Ti-6Al-4V microstructure where the white dotted lines show the grain boundary α and the white arrows examples of α laths.

2. Method

In this work, the microstructure in material built from five different AM processes have been characterised, regarding the prior β grain size, α lath thickness and grain boundary α. The chosen AM processes for this investigation are LMwD, where a laser is used as energy source and a wire for deposition of material. Two differently built batches of material from this AM process have been investigated, one in which each layer was deposited continuously (LMwD-0) and one which included a two minutes break between each added layer (LMwD-2). Laser Metal powder Deposition (LMpD) is also investigated here, which is an AM process where laser combined with blown powder is used. TIG-torch energy source combined with wire is yet another process that has also been included in this work and it is here called Shaped Metal Deposition (SMD). These three AM processes (LMwD, LMpD and SMD) belong to the directed energy deposition group of AM processes. Two powder bed fusion processes were also investigated - Electron Beam Melting (EBM), where an electron beam is used as energy source, and Selective Laser Melting (SLM), where a laser beam energy source is used.
The SMD sample received a post heat treatment of 670 °C/2h, while the LMwD samples were exposed to 704 °C/2h. These heat treatments were performed for releasing stresses and do not affect the microstructure [16]. For the LMpD sample no post heat treatment was performed. The powder bed fusion materials were built using an EOS 290 machine (SLM) and an Arcam Q20 machine (EBM). None of the powder bed fusion materials where post build treated. In this work cross-sections perpendicular to the layers were characterised, i.e. cross sections across many layers. The sample

preparation was done according to conventional processes for titanium alloys. To reveal the microstructure, the samples were etched by Kroll's etchant consisting of 92 ml distilled water, 6 ml HNO_3 and 2 ml HF. The microstructure characterisation was done using a light optical microscope (LOM, Nikon Eclipse MA200) and its software (NIS Elements BR) and a stereomicroscope (Nikon SMZ1270).

The prior β grain measurements were performed in accordance to ASTM 112-13 [17] and by using large image stitching in the LOM. However, in order to measure the prior β grain size accurately, before each measurement the prior β grain boundaries were carefully marked in the images at a higher magnification by using an image editing software (Photoshop CC 2015), with stereomicroscope images taken into consideration. In total 500 prior β grain size measurements were done per sample.

The grain boundary α thickness was measured by using a grid tool in the LOM to avoid biased results and to obtain a more accurate average value of the thickness. A grid of squares (each square: 100x100 μm) was applied in the live image and within squares having grain boundary α one measurement was conducted. In total 200 measurements per sample were carried out.

The α laths were measured in five different areas on each sample; see illustrated location of the measurements in figure 2, denoted A-E in the EBM sample. In total 200 measurements per area were conducted, i.e. 1000 measurements per sample.

3. Results and discussion

3.1 Prior beta grains

In figure 2 macroscopic LOM overview images of Ti-6Al-4V material built with the five different AM processes are shown. Two characteristic prior β grains have been indicated with white dotted lines in the LMwD-0 and SLM materials. In Figure 3 the result of the prior β grain size measurements are summarized. Here it can be seen that material from the powder bed fusion processes (EBM and SLM) have smaller prior β grain size than the material built using directed energy deposition processes (LMwD, LMpD and SMD). This is clear also when comparing the materials of EBM and SLM with e.g. LMwD and SMD in Figure 2.

The prior β grain size is mainly determined by the time the material is exposed to temperatures above the β transus temperature. Normally the β transus temperature is around 995°C for Ti-6Al-4V, but the β transus temperature depends on the exact chemical composition of the alloy and can therefore vary between 970~1000°C [18]. Longer time and increased temperature above the β transus temperature lead to increased β grain size. Based on the findings in the present investigation it can be seen that the material built with the LMwD-0 and SMD processes have the largest prior β grains, whereas the materials from both powder bed fusion processes (EBM and SLM) show significantly smaller prior β grain size. It is thus evident that the materials built from LMwD-0 and SMD endure longer time above the β transus temperature compared to EBM and SLM. The reason for these differences are found in the different process setups (powder bed versus wire based deposition) and their distinct process parameters that influence the heat input. These differences then determine the cooling rate from above the β transus temperature and thus the prior β grain size.

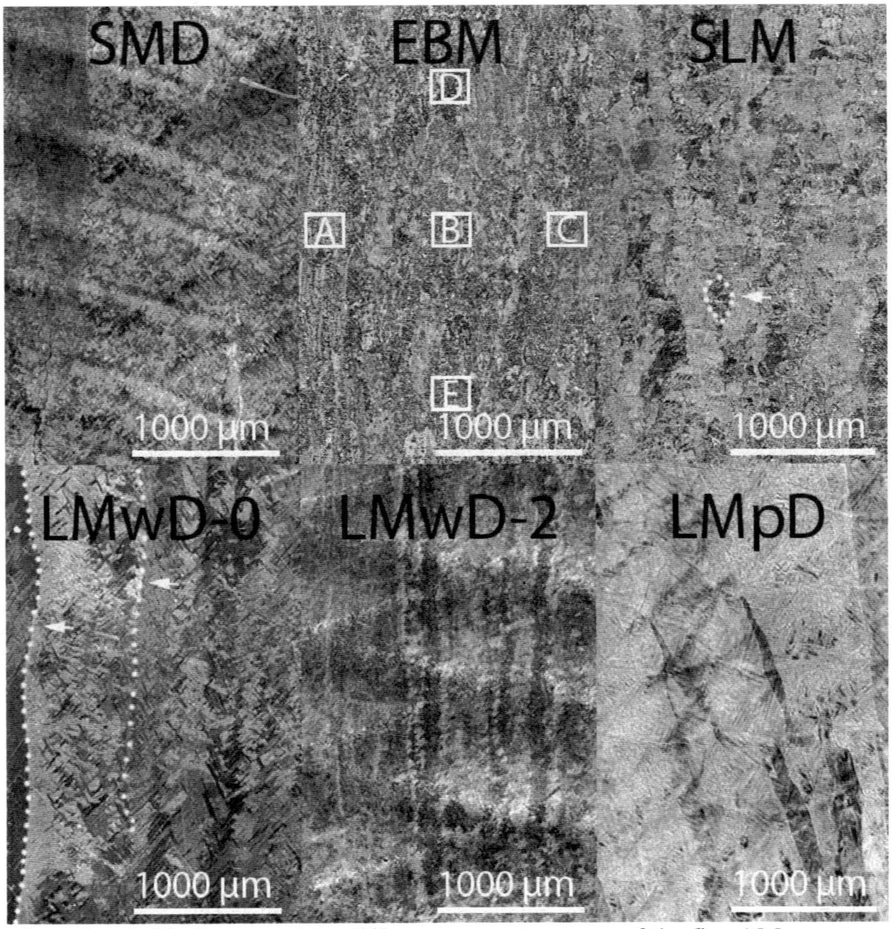

Figure 2. Overview LOM images of the different macrostructures of the five AM processes. The white arrows show prior β grain boundaries (white dotted lines) and the letters A-E illustrate the location of the α lath measurements on all the samples.

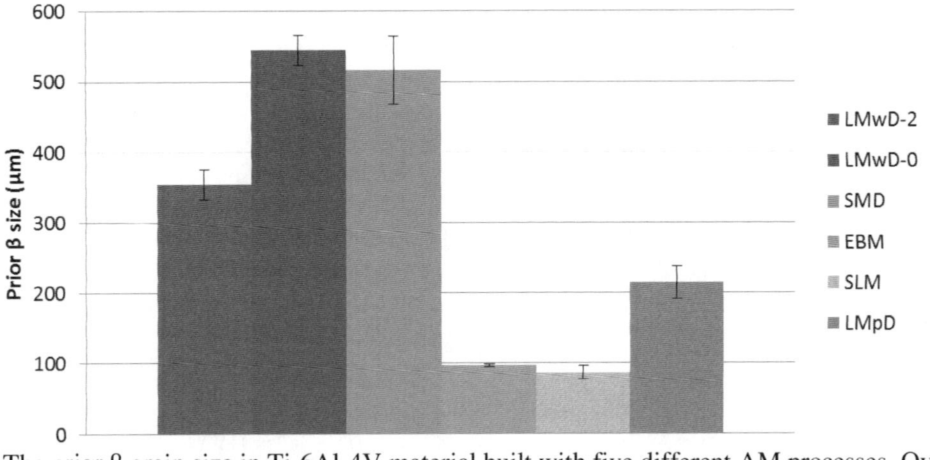

Figure 3. The prior β grain size in Ti-6Al-4V material built with five different AM processes. Overall the directed energy deposition based AM processes (LMwD, LMpD and SMD) have larger prior β grain sizes compared with material from the powder bed fusion processes (EBM and SLM).

3.2 Grain boundary alpha

The result of the grain boundary α measurements can be seen in Table **1**. The LMwD-0 material exhibits the thickest grain boundary α, followed by material from the powder bed fusion processes (EBM and SLM). Materials from LMwD-2 and SMD processes exhibit discontinuous grain boundary α with the average thickness less than one micrometer. Material from the LMpD process contains no grain boundary α and this is a result of the fast cooling of this AM process.

Table 1. Average and maximum thickness of grain boundary α in Ti-6Al-4V material built with different AM processes.

Name	Avg. grain boundary α	Max grain boundary α
LMwD-2	0.7 (discontinuous) ± 0.1	1.2
LMwD-0	3.5 ± 1.4	7.7
LMpD	-	-
SMD	0.6 (discontinuous) ± 0.1	1.0
EBM	2.9 ± 0.5	4.3
SLM	2.6 ± 0.7	4.6

3.3 Parallel bands

In Figure **2** parallel bands can be seen in the SMD, LMwD-2 and LMpD material. These are characterised by their wavy texture, which can be compared to e.g. EBM and SLM that have no parallel bands. Others have reported parallel bands for LMpD [13], LMwD [15] and SMD [4] as well. In the present work, the regions in between bands contain thicker α laths than regions within a band, where thinner α laths dominate. The reason for this is that the region with coarser α laths is closer to the fusion zone of next deposited layer, which renders in some coarseing of the microstructure closest to the fusion zone. This corresponds well with the findings of Sandgren et al. [2].

3.4 Alpha laths

Figure 4 shows overview LOM images of the different microstructures found in material from the investigated AM processes. The microstructure in material built using the LMpD process is unclear and difficult to distinguish, which suggests a martensitic transformed microstructure. The microstructures in the SMD and LMwD-2 material are very similar and are basket weave types of microstructures. The same type of microstructure, but coarser, is also found in the powder bed fusion manufactured materials, i.e. for EBM and SLM. The microstructure in the LMwD-0 material shows colonies of α laths, i.e. several α laths oriented in parallel forming colonies. This indicates that the cooling rate below the β transus temperature was slower than in materials built with the other AM processes, and the microstructure of LMwD-0 equates to the Widmanstätten colony morphology. In figure 4 the thicknesses of the laths are clearly visible, except for the LMpD material. The appearance of the α lath thicknesses is supported by the measurements shown in Figure 5 where SMD has the thinnest α laths with an average thickness of 0.7 μm, followed by LMwD-2 (1 μm), SLM (1.5 μm), EBM (1.5 μm) and LMwD-0 (2.3 μm). Seifi et al. [6] investigated EBM Ti-6Al-4V material and found the α laths to be in the range of 1 to 1.2 μm. Rafi et al. [14] investigated the α laths of SLM Ti-6Al-4V material and determined the thickness to be in the range of 1 to 2 μm. These results correspond well with the results in the present work.

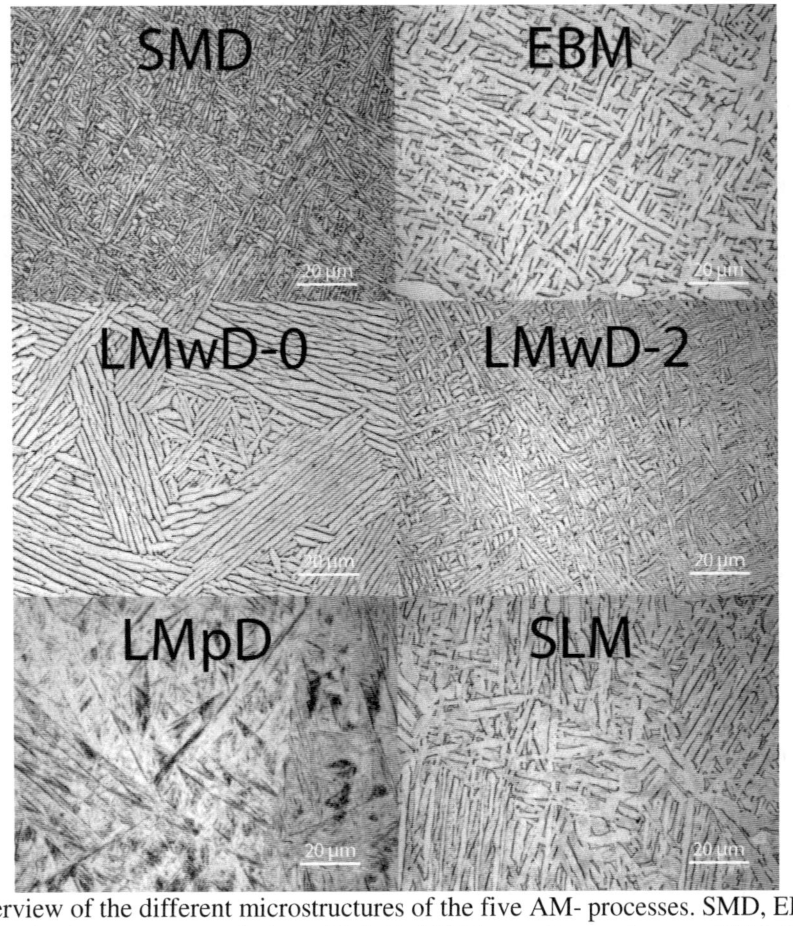

Figure 4. Overview of the different microstructures of the five AM- processes. SMD, EBM, LMwD-2 and SLM have basket weave morphology, LMwD-0 Widmanstätten colony and LMpD martensitic morphology.

Sandgren et al. [2] measured the α lath thickness to be 0.7 µm for their LMwD Ti-6Al-4V material, which corresponds well with the findings of Baufeld et al. [12], that found the corresponding thickness to be 0.6 µm. These results are close to the results found in the present work in the LMwD-2 Ti-6Al-4V material (1 µm). In order to investigate if the size of α laths varies in different regions of AM materials, different regions were selected within the same AM material, as depicted schematically in the micrograph for the EBM material in figure 2. The results from these measurements, shown in figure 5, reveal that in most types of AM material the α lath thickness is quite constant, even though the standard deviation for some of the materials is large, especially for the LMwD-0 material.

From the microstructural characterisation, it is possible to get an estimation of the mechanical properties of the different material. The α colony size is known to be the microstructural feature that is most important for the mechanical properties as it correlates with the slip length of the material [19]. Smaller α colonies render in higher strength and their size is somewhat correlating with the α lath and grain boundary α thickness, where the prior β grain set the boundary for how large the α colonies can become. Faster cooling rates decrease the size of the α colonies along with the thickness of the α laths and grain boundary α i.e. thinner α lath is closely linked to smaller α colonies. So, from the findings LMpD is estimated to have the highest strength of the here investigated material, while LMwD-0 is estimated to have the lowest.

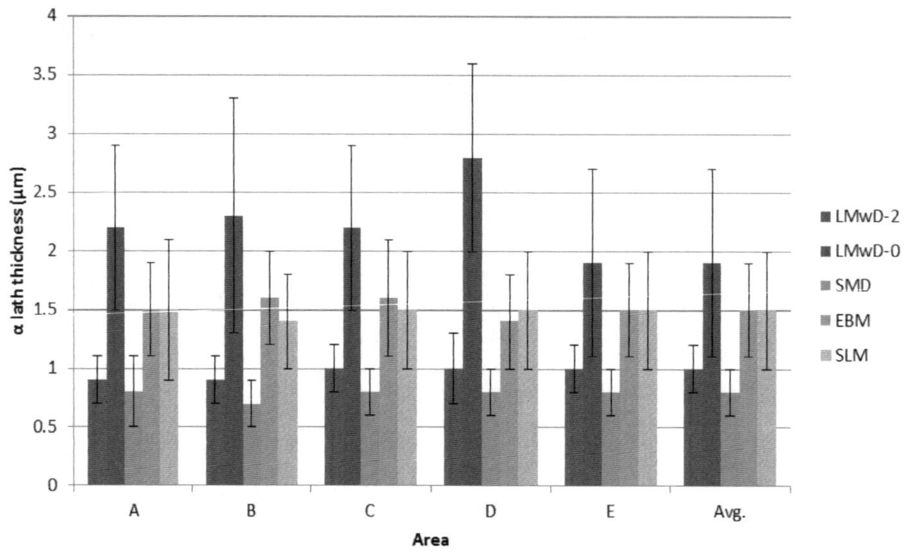

Figure 5. The α lath thickness in micrometer of the different AM processes on the different areas A-E, from figure 2. The column Avg. is the average of the five areas A-E.

The thickness of the grain boundary α along with the α colony size affect the ductility of the material, where very high cooling rates i.e. small α colonies and thin/non-existent grain boundary α render in poor ductility. Very slow cooling rates do not lead to the best ductility either, the ε_{max} is rather found at cooling rates of 100° C/min [19]. The cooling rates of the here investigated materials are above this rate which consequently results in that LMwD-0 is estimated to be the material with the highest ductility while LMpD should have the lowest. In between these two materials the mechanical properties of the other materials are most likely found by following the α lath thickness trend presented in this work.

4. Conclusions

In this work, materials manufactured with the powder bed fusion processes (EBM and SLM) have similar microstructures, i.e. prior β grain size, grain boundary α, and α lath size. The directed energy deposition processes (LMwD, LMpD and SMD) render in larger prior β grains than in material from the powder bed fusion processes. The parallel band phenomenon was present in materials from the SMD, LMpD and LMwD-2 processes, while non-existing in material from the LMwD-0, EBM and SLM built materials. Continuous manufacturing of Ti-6Al-4V material with the LMwD process, compared with introducing a 2 minutes holding time between each deposited layer, renders in significantly larger prior β grain size, thicker α phase layers in the prior β grain boundaries and thicker α laths.

Acknowledgements

The author would like to thank GKN Aerospace Engine Systems AB for their support throughout the project. Furthermore, the authors would also like to thank the student Viktor Sandell for his contributions.

References

[1] Śliwa R E, Bernaczek J and Budzik G 2016 The application of direct metal laser sintering (DMLS) of titanium alloy powder in fabricating components of aircraft structures *Key Engineering Materials* **687** pp 199-205

[2] Sandgren H R, Zhai Y, Lados D A, Shade P A, Schuren J C, Groeber M A, Kenesei P and Gavras A G 2016 Characterization of fatigue crack growth behavior in LENS fabricated Ti-6Al-

4V using high-energy synchrotron x-ray microtomography, *Additive Manufacturing A* **12** pp 132-41
[3] Mandil G, Le T T, Paris H and Suard M 2016 Building new entities from existing titanium part by electron beam melting: microstructures and mechanical properties *The International Journal of Advanced Manufacturing Technology* **85** pp 1835–46
[4] Baufeld B, van der Biest O and Gault R 2009 Microstructure of Ti-6Al-4V specimens produced by shaped metal deposition *International Journal of Material Research* **100** no. 11 pp 1536-42
[5] Al-Bermani S, Blackmore M L, Zhang W and Todd I 2010 The origin of microstructural, texture, and mechanical properties in electron beam melted Ti-6Al-4V *Metallurgical and Materials Transactions A* **41** pp 3422-34
[6] Seifi M, Salem A, Satko D, Shaffer J and Lewandowski J J 2016 Defect distribution and microstructure heterogeneity effects on fracture resistance and fatigue behavior of EBM Ti–6Al–4V *International Journal of Fatigue* **94** pp 263–87
[7] Sridharan N, Chaudhary A, Nandwana P and Babu S S, 2016, Texture evolution during laser direct metal deposition of Ti-6Al-4V *Journal of the Minerals, Metals & Materials Society* **68** no. 3 pp 772-7
[8] Sefer B, Roa J J, Mateo A, Pederson R and Antti M-L 2016 Evaluation of bulk and alpha-layer properties in Ti-6Al-4V at micro- and nano-metric length scale *Proc. of the 13th World Conference on Titanium* pp1619-24 DOI: 10.1002/9781119296126.ch271
[9] Gibson I, Rosen D and Stucker B 2010 *Additive Manufacturing Technologies: Rapid prototyping to direct digital manufacturing* p 246 Springer
[10] Formanoir C, Michotte S, Olivierrigo L G and Godet S, 2016, Electron beam melted Ti–6Al–4V: Microstructure,texture and mechanical behavior of the as-built and heat-treated material, *Materials Science & Engineering A* **652** pp 105–19
[11] Wang F, Williams S, Colegrove P and Antonysamy A A 2012 Microstructure and mechanical properties of wire and arc additive manufactured Ti-6Al-4V *Metallurgical and Materials Transactions A* **44** pp 968-77
[12] Baufeld B, Brandl E and Biesta O 2011 Wire based additive layer manufacturing: Comparison of microstructure and mechanical properties of Ti–6Al–4V components fabricated by laser-beam deposition and shaped metal deposition *Journal of Materials Processing Technology* **211** no. 6 pp 1146–58
[13] Zhai Y, Lados D A, Brown E J and Vigilante G N 2016 Fatigue crack growth behavior and microstructural mechanisms in Ti-6Al-4V manufactured by laser engineered net shaping *International Journal of Fatigue* **93** pp 51-63
[14] Rafi H, Karthik N, Gong H, Starr T L and Stucker B E 2013 Microstructures and mechanical properties of Ti6Al4V parts fabricated by selective laser melting and electron beam melting *Journal of Materials Engineering and Performance* **22** no. 12 pp 3872–83
[15] Kelly S and Kampe S 2004 Microstructural evolution in laser-deposited multilayer Ti-6Al-4V builds: Part I. microstructural characterization *Metallurgical and Materials Transactions A* **35** pp 1861-67
[16] Vrancken S, Thijs L, Kruth J and Van Humbeeck J 2012 Heat treatment of Ti6Al4V produced by Selective Laser Melting: Microstructure and mechanical properties *J.Alloys Compounds* **541** pp 177-85
[17] ASTM international *Standard test methods for determining average grain size* E112 – 13.
[18] Froes F H 2015 *Titanium Physical Metallurgy Processing and Applications* ASM International p 228
[19] Lütjering G 1998 Influence of processing on microstructure and mechanical properties of (α β) titanium alloys *Materials Science and Engineering: A* **243** no. 1 pp 32-45

IOP Conf. Series: Materials Science and Engineering **258** (2017) 012008 doi:10.1088/1757-899X/258/1/012008

Evolution of the microstructure in carbon nanotube reinforced Nickel matrix composites processed by high-pressure torsion

K Aristizabal[1,*], S Suárez[1], A Katzensteiner[2], A Bachmaier[2], and F Mücklich [1, 3]

[1] Chair of Functional Materials. Department of Materials Science, Campus D3.3, 66123 Saarbrücken, Germany.

[2] Erich Schmid Institute, Austrian Academy of Sciences, Jahnstrasse 12, 8700 Leoben, Austria

[3] Materials Engineering Center Saarland, Campus D3.3, D-66123, Saarbrücken, Germany.

E-mail: katherine.aristizabal@uni-saarland.de, s.suarez@mx.uni-saarland.de, andreas.katzensteiner@oeaw.ac.at, andrea.bachmaier@oeaw.ac.at, muecke@matsci.uni-sb.de

Abstract. Carbon nanotube (CNT)-reinforced nickel matrix composites were processed using high-pressure torsion (HPT) at room temperature (RT). Different CNT weight fractions were used in order to study the behavior of the composites in the "as sintered" and the "as deformed" conditions and to determine the effect of the amounts of CNT added on the different processing methods. The samples were analyzed by means of Vickers microhardness and electron microscopy. According to the results, increasing the CNT content in the "as sintered" condition increases the agglomerate size but decreases only slightly the grain size. Regarding the "as deformed" condition it showed little to negligible effect in further refining the microstructure. By means of HPT the hardness was increased up to 800%. It was concluded that the microstructure could be further improved in terms of grain size and agglomerate size and distribution by means of HPT.

1. Introduction

Seeking the stabilization of the microstructure against grain growth is critical in nanocrystalline (NC) and ultrafine-grained (UFG) materials processed by severe plastic deformation (SPD), since they possess high amounts of stored energy in their large grain boundary area, giving them an unstable nature. This instability can activate microstructural recovery and grain growth processes even at low temperatures that can alter the material properties considerably. In some studies, the addition of reinforcing phases to metallic matrices subjected to SPD has been performed with the aim of overcoming this issue and given a proper distribution of the particles [1] they even result in enhanced mechanical properties [2, 3]. The resulting metal matrix composites (MMC) behave differently than pure metals and provided that the appropriate reinforcing phase is chosen and a uniform distribution of the agglomerates can be obtained, the thermal instability drawback might be overcome by making use of different mechanisms (e.g. by hindering the grain boundary and dislocation mobility).

IOP Conf. Series: Materials Science and Engineering **258** (2017) 012008 doi:10.1088/1757-899X/258/1/012008

In this study, we propose the use of carbon nanotubes as reinforcing phase in Ni matrix composites, acting a stabilizing second phase. The choice is based on the fact that, though they possess outstanding intrinsic properties, they have also shown the ability to pin the grain boundaries under thermal inputs [4, 5]. The results presented here are emphasized on the comparison between the "as sintered" and the "as deformed" conditions of the studied composites and serve as a starting point of a deeper microstructural study, which will be discussed elsewhere.

2. Experimental

Carbon nanotube (CNT)-reinforced nickel matrix composites were processed using high-pressure torsion (HPT). Different CNT weight fractions were studied, namely: 0.5 wt.% (2.4 vol.%), 1 wt.% (4.8 vol.%), 2 wt.% (9.5 vol.%) and 3 wt.% (14.5 vol.%). These composites were sintered by means of hot uniaxial pressing (HUP) and afterwards subjected to HPT at room temperature (RT) using different number of turns (1T, 4T, 10T and 20T) under 4 GPa of pressure.

The "as sintered" condition was analyzed by means of Vickers micro-hardness, scanning electron microscopy (SEM), with a resolution of 125 nm per pixel, and electron backscattered diffraction (EBSD). The electron microscopy characterization was performed on a dual beam system Helios NanoLab 600 (FEI). The EBSD measurements were performed with an EDAX TSL detector, with a 20 kV voltage bias and 22 nA current. The step size for the analysis was varied for the different samples, ranging from 150 to 500 nm.

The evolution of the microstructure along the radial direction, corresponding to increasing equivalent strain values ε_{vM} [6] (according to $\varepsilon_{vM} = \frac{2\pi nr}{t\sqrt{3}}$, where *n* is the number of turns, *t* is the sample thickness and *r* the distance from the center of the sample), was studied by means of SEM using backscattered electron imaging (BSE) as a faster way of obtaining information about the mean grain size (equivalent diameter), given that in the "as deformed" samples, the significant residual stresses made difficult the proper sample preparation for EBSD. The shape, size and distribution of the CNT agglomerates were also studied. The images were taken every 1 mm along the radial direction with a resolution of 2.5 nm per pixel.

Vickers micro-hardness, which is extensively used as a quick way of assessing the saturation in the microstructural refinement of such composites, was performed every 0.5 mm along the radial direction, using an indenter Struers DuraScan 50/70/80.

3. Results and discussion

The mean grain size values obtained for the different compositions of the HUP samples by means of EBSD lay all within the micrometer range and are summarized in Table 1. From these values, no clear correlation between the mean grain size and the CNT concentration can be extracted. This might be due to an incomplete pinning effect of the CNTs on the grain boundaries during the densification stage [7].

Table 1. Mean grain sizes of HUP samples obtained by EBSD. For the determination of the mean and standard deviation, a log-normal distribution was assumed.

HUP Samples	Mean grain size [μm]
0.5wt.% MWCNT/Ni	4.80 ± 2.89
1wt.% MWCNT/Ni	3.10 ± 1.17
2wt.% MWCNT/Ni	3.42 ± 1.55
3wt.% MWCNT/Ni	3.67 ± 1.67

IOP Conf. Series: Materials Science and Engineering **258** (2017) 012008 doi:10.1088/1757-899X/258/1/012008

Figure 1 shows the evolution of the grain size (equivalent diameter) along the radius of the samples, obtained from BSE images for the deformed samples. Here, it was assumed that the grains are the areas delimited by the different gray values within the image, and the misorientation angle between the grains, which cannot be measured with this method, was not considered. The results show a large decrease in grain size after HPT and at early stages of the deformation, but only a subtle decrease with increasing equivalent strains reaching values down to 100 nm, which is independent of the CNT volume fraction. From this, it can be stated that the increment of the CNTs volume fraction has little to negligible effect on further refining the microstructure during deformation, presenting similar results regarding the grain size for the same processing conditions (as shown in Figure 1).

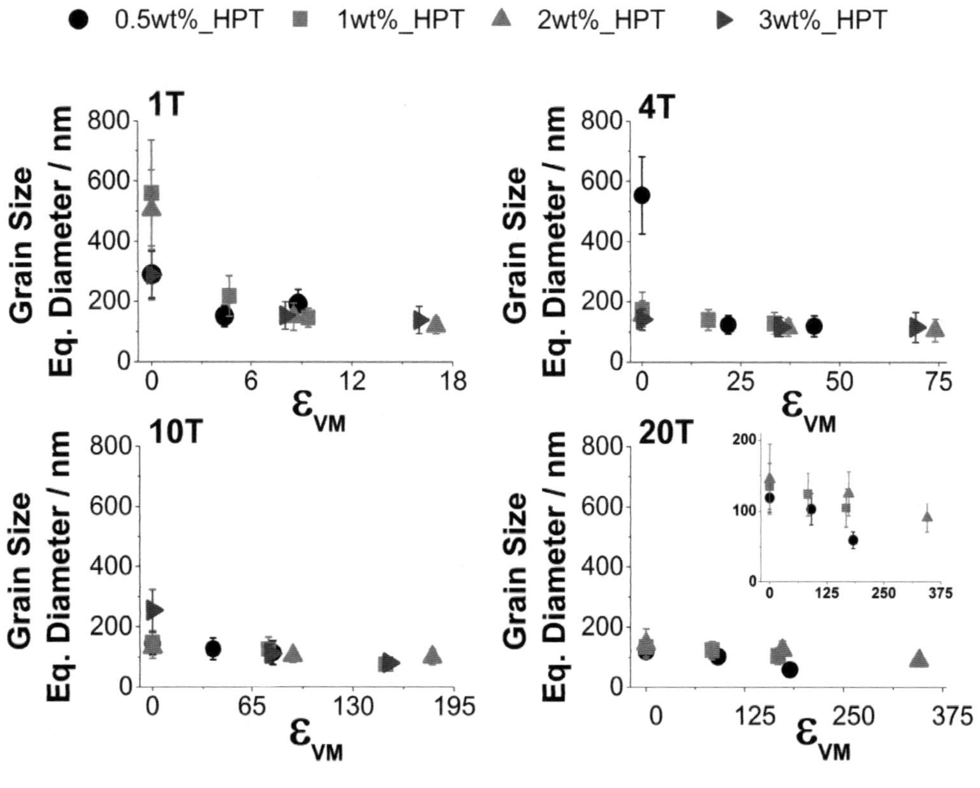

Figure 1. Change in grain size (equivalent diameter) with increasing equivalent strains. (Sample 3wt.% could not be further deformed by HPT at RT due to its high hardness causing slippage between the sample and the anvils).

Regarding the hardness, table 2 summarizes the values for the as sintered condition (HUP). The hardness increases only slightly with increasing CNT content. This correlates well to what has been previously reported in the literature [4], and is traced back to the role played by the partial CNT concentration in the dislocation mobility hindering in these composites during the indentation stage. The CNTs act as pinning points to the dislocation fronts, progressively reducing their energy. This effect combined with the microstructural refinement, leads to a combination of two strengthening mechanisms, namely: grain boundary strengthening (Hall-Petch effect) and dispersion strengthening (Orowan mechanism). Since there are no relevant changes in the mean grain size in the HUP state, the largest role in the enhancement might thus be provided by an Orowan-type mechanism.

IOP Conf. Series: Materials Science and Engineering **258** (2017) 012008 doi:10.1088/1757-899X/258/1/012008

Table 2. Vickers hardness HUP samples as a function of the partial CNT fraction.

HUP Samples	Hardness [GPa]
0.5wt.% MWCNT/Ni	1.17 ± 0.05
1wt.% MWCNT/Ni	1.17 ± 0.03
2wt.% MWCNT/Ni	1.20 ± 0.07
3wt.% MWCNT/Ni	1.31 ± 0.07

In Figure 2 the evolution of the hardness along the radial direction is shown for the HPT samples. The results show a significant increase in hardness with increasing equivalent strains of about 800 % with respect to the as sintered condition. Interestingly, even after 20 turns the hardness values continue to increase indicating that the onset of the microstructural refinement saturation has not been reached for the equivalent deformation here studied. Remarkably, all the analyzed concentrations show the similar trends in the hardening behavior. Thus, it could be stated that this feature is mainly based on the nature of the matrix and cannot be straightforwardly correlated to the amount of second phase. Nevertheless, a slight decrease in the "slope" of the "lines" is observed for increasing CNT content, which can be a "sign" that the dislocation and grain boundary mobility are hindered by the CNT during deformation.

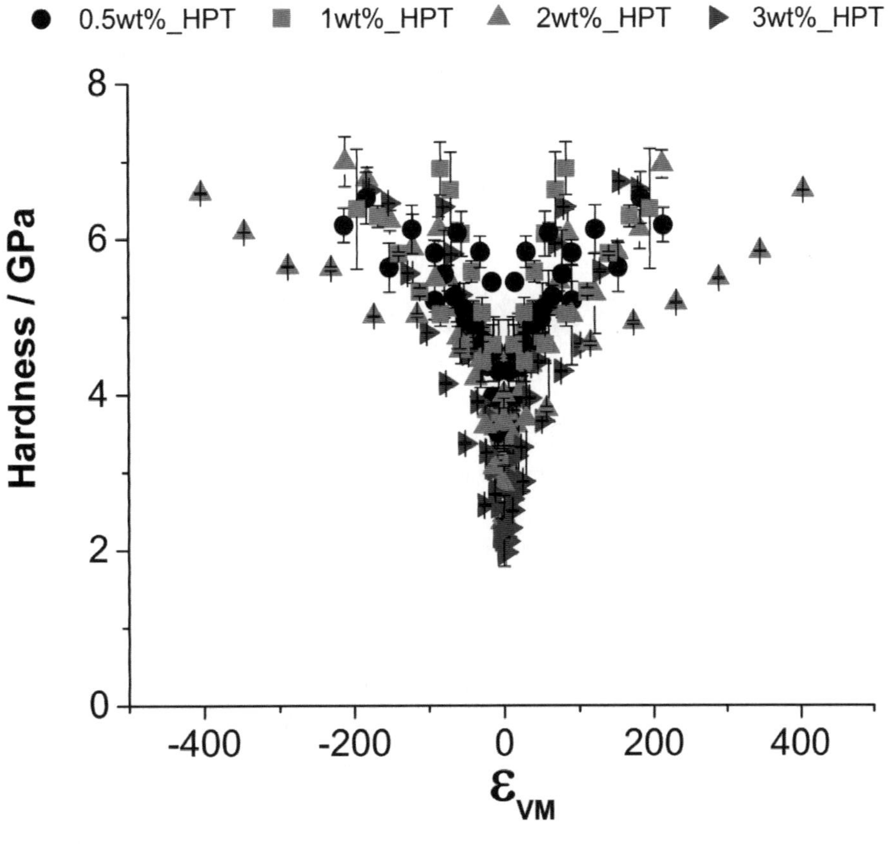

Figure 2. Variation of the hardness with increasing equivalent strains. The negative values of the equivalent strain represent only the opposite direction of the radius and do not depict a physical value.

Focusing on the effect of the deformation on the CNT agglomerate morphological characteristics, the analysis is directed to the study of the change in size and shape of the CNT clusters. Figure 3 shows binarised images for the HUP samples showing the CNT agglomerates and the corresponding mean agglomerate size values are summarized in table 3. From this observation, it becomes evident that the agglomerate size increases with increasing CNT content. This is explained by the fact that for larger CNT concentrations, there is a lower mean free path for CNT re-clustering. This has been already observed in sintered bulk metal matrix composites, and is traced back to the fact that, since the CNTs are not embedded in the metal particles, the material transport during sintering brings them closer together increasing their probability of reagglomeration due to van der Waals interactions [8].

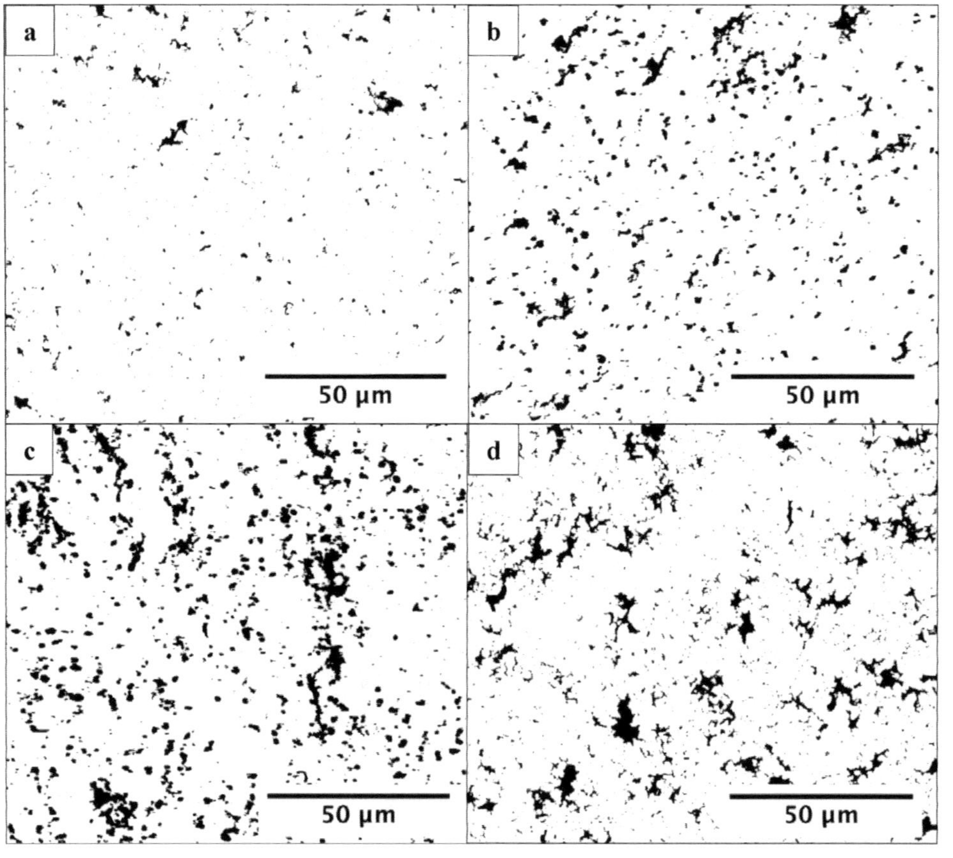

Figure 3. Binary images HUP samples. a) 0.5wt.% MWCNT/Ni, b) 1wt.% MWCNT/Ni, c) 2wt.% MWCNT/Ni and d) 3wt.% MWCNT/Ni.

IOP Conf. Series: Materials Science and Engineering **258** (2017) 012008 doi:10.1088/1757-899X/258/1/012008

Table 3. Mean agglomerate sizes of HUP samples obtained by image analysis using the image-analysis software FIJI.

HUP Samples	Mean agglomerate size [µm]
0.5wt.% MWCNT/Ni	1.91 ± 1.52
1wt.% MWCNT/Ni	2.44 ± 1.62
2wt.% MWCNT/Ni	3.11 ± 2.41
3wt.% MWCNT/Ni	3.77 ± 2.34

Concerning the as-deformed state, figure 4 presents BSE images taken at 3 mm from the center for different compositions and number of turns. The evolution of the shape, the distribution and the agglomerate size can be exemplarily visualized in this figure. The agglomerates develop an elongated shape in the shear direction at early stages of the deformation. With higher number of turns, the agglomerates become more rounded and their distribution improves. The agglomerate size decreases significantly when comparing to the as sintered condition, reaching values down to the sub-micrometer range.

Weight / Turns **1** **4** **10** **20**
Percent

0.5

1

2

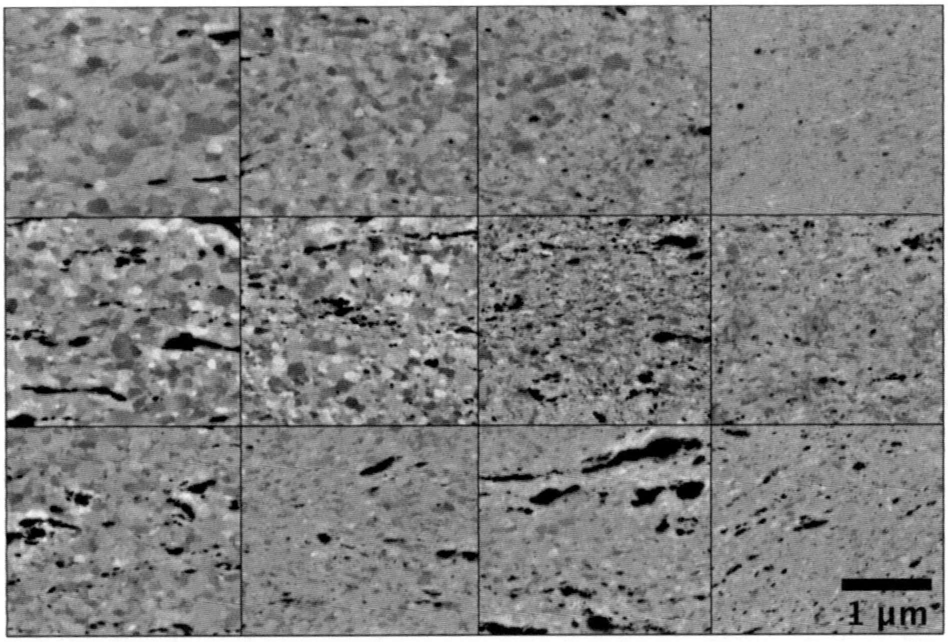

Figure 4. BSE images for a sample set at a radius of 3 mm. (Equivalent strain increases from left to right).

The decrease in grain size is within the same range and remains almost constant for the higher equivalent strains (as shown in figure 2), resembling the saturation onset of the microstructural refinement. Nevertheless, according to the Vickers hardness results, this statement cannot be confirmed for the parameters of HPT here studied. The results suggest that the microstructure can be further refined in terms of grain size (see inset in figure 1) and improved in terms of agglomerate size and distribution by means of HPT.

4. Concluding remarks

CNT reinforced Ni matrix composites with different CNT contents were processed by HPT at RT using 4 GPa of pressure. When comparing to the HUP samples, it could be shown that after HPT not only the grain size but also the agglomerate size decreases and the microhardness increases significantly. Nevertheless, increasing the CNT content both in HUP and in HPT samples showed little to negligible effect on further refining the microstructure. Furthermore, the size and spatial distribution of the agglomerates was improved by HPT. Finally, the results suggest that the microstructure can be further improved in terms of grain size and agglomerate size and distribution by means of HPT (e.g. by optimizing the parameters used during processing).

Acknowledgments

K. Aristizabal would like to thank the German Academic Exchange Service (DAAD) for financial support. S. Suarez and K. Aristizabal gratefully acknowledge the financial support from DFG (Grant: SU911/1-1). A. Katzensteiner and A. Bachmaier gratefully acknowledge the financial support by the Austrian Science Fund (FWF): I2294-N36. This project has received funding from the *European Union's Horizon 2020 research and innovation programme* under grant agreement No 644013 (CREATe-Network).

References

[1] Bachmaier A and Pippan R 2013 *Int. Mater. Rev.* 58 41–62
[2] Islamgaliev R K, Buchgraber W, Kolobov Y R, Amirkhanov N M, Sergueeva A V, Ivanov K V and Grabovetskaya G P 2001 *Mater. Sci. Eng. A* 319–21 872–6
[3] Bachmaier A and Pippan R 2011 *Mater. Sci. Eng. A* 528 7589–95
[4] Suarez S, Lasserre F, and Mücklich F 2015 *Mater. Sci. Eng. A* 587 381–386
[5] Suarez S, Lasserre F, Soldera F, Pippan R and Mücklich F 2015 *Mater. Sci. Eng. A* 626 122–7
[6] Stüwe H P 2003 *Adv. Eng. Mater.* 5, 291–5
[7] Reinert L, Suarez S, Müller T, and Mücklich F 2017 *Adv. Eng. Mater.* DOI: 10.1002/adem.201600750
[8] Rossi P, Suarez S, Soldera F, and Mücklich F. 2015 *Adv. Eng. Mater.* 17(7) 1017-21.

IOP Conf. Series: Materials Science and Engineering **258** (2017) 012009 doi:10.1088/1757-899X/258/1/012009

Effect of surface etching and electrodeposition of copper on nitinol

E Ramos-Moore[a,b,*], A Rosenkranz[a, b], L F Matamala[c], A Videla[c,b], A Durán[d] and J Ramos-Grez[e,b]

[a] Instituto de Física, Pontificia Universidad Católica de Chile, Santiago 7820436, Chile.
[b] Centro de Investigación en Nanotecnología y Materiales Avanzados "CIEN-UC", Pontificia Universidad Católica de Chile, Santiago, Chile.
[c] Departamento de Ingeniería en Minas, Pontificia Universidad Católica de Chile, Santiago, Chile.
[d] Departamento de Ingeniería Metalúrgica, Universidad de Santiago de Chile, Avda. Libertador Bernardo O´Higgins 3363, Casilla 10233, Santiago, RM, Chile
[e] Departamento de Ingeniería Mecánica y Metalúrgica, Pontificia Universidad Católica de Chile, Santiago, Chile.

*evramos@uc.cl

Abstract. Nitinol-based materials are very promising for medical and dental applications since those materials can combine shape memory, corrosion resistance, biocompatibility and antibacterial properties. In particular, surface modifications and coating deposition can be used to tailor and to unify those properties. We report preliminary results on the study of the effect of surface etching and electrodeposition of Copper on Nitinol using optical, chemical and thermal techniques. The results show that surface etching enhances the surface roughness of Nitinol, induces the formation of Copper-based compounds at the Nitinol-Copper interface, reduces the austenitic-martensitic transformations enthalpies and reduces the Copper coating roughness. Further studies are needed in order to highlight the influence of the electrodeposited Copper on the memory shape properties of NiTi.

1. Introduction

Shape memory alloys (SMA) are known for a long time since the first observation of this effect was made in 1932 [1]. After that, intense fundamental research on this effect made it possible to generate commercial applications [2]. One of the most prominent and extensively studied SMA is Nickel-Titanium (NiTi), also known as Nitinol, which stands for Nickel Titanium Naval Ordnance Laboratory. Nitinol represents a highly ordered intermetallic phase with a nearly quasi equiatomic composition of 50 % Ni and 50 % Ti. Due to the fact that the shape memory effect can just be found in a very narrow concentration range, different alloying elements are typically used in order to increase this range and to enhance certain physical properties [3]. In this context, Niobium can be added to NiTi to generate an alloy with a wide hysteresis thus leading to a more pronounced two-way shape memory effect as well as better damping properties [4]. Copper can be used to narrow the transformation hysteresis, thus inducing high frequency response [5] and making it suitable for micromechanical actuators. In terms of applications, NiTi-alloys are very interesting for the medical and dental sector since those alloys combine shape memory as well as pseudoelasticity with remarkable material properties such as good biocompatibility [6], corrosion resistance and good mechanical properties [7].

Despite the proved biocompatibility of NiTi [6], major concerns related to medical issues arise based upon the chemical composition and the high Ni content. It is well-known that Ni is highly toxic for the human body and it can be assumed that these concerns might decelerate the application of this alloy in the medical sector [7]. From the material's point of view, the Ni release can be directly correlated to the corrosion resistance. Consequently, various surface treatments such as surface modifications or coatings can be thought of to improve and tailor the corrosion resistance as well as the biocompatibility [8,9]. Chemical and electrochemical methods such as electropolishing and chemical etching need to be considered since those methods can be utilized to avoid defective surface layers and oxidation of the surface [6]. Although antibacterial metals, such as Silver and Copper, can be toxic to the human body when its concentration exceeds certain threshold, the control of their adherence onto NiTi during coating deposition opens the possibility to improve the functionality of SMAs by adding the antibacterial property to the material surface.

In the context of medical or dental applications, two other aspects, namely biocompatibility and antibaterical properties, need to be addressed. An easy way to enhance the biocompatibility of SMA is to use a biologically active coating [6,7,10]. Regarding the antibaterical properties, different attempts have been made in order to induce an antibacterial function on SMA. Ahmed et al. [11] produced a biocomposite of gold nanoparticles and chitosan (natural polymer). This biocomposite was electro-deposited on NiTi to investigate the effect of this coating on biocompatibility and antibacterial effects. They could verify that this coating reduces the Ni release by a factor of 20. Additionally, due to the presence of the gold nanoparticles, an antibacterial effect could be observed. Another interesting approach to improve the antibacterial effect was the addition of Silver [12], but to the best of our knowledge, no systematic study on the electrodeposition of Copper on NiTi has being performed.

In this work, we study topographical, chemical and phase transformation effects of surface etching as well as Copper electrodeposition on Nitinol. Our hypothesis claims that etching may increase the surface area as well as the surface reactivity of Nitinol in order to improve the nucleation and growth of crystalline Copper coatings. Our results show interesting properties of the etching process as a way to further enhance the integration of Copper into the NiTi surface [13-15] thus combining shape memory and antibacterial properties.

2. Experimental Methods

Nitinol samples (Nitinol UNSPSC 30000000) were obtained from Kellogg´s Research Labs in form of plates (1 mm thick, 45 mm wide and 150 mm long). The composition of the samples was 50 at.-% Ni and 50 at.-% Ti. The etching treatment consisted in exposing one of the Nitinol sample face to a 10% HNO_3 solution during 24 hours at room temperature and no agitation. Atomic Force Microscopy (AFM), NanoWizard 3 Bioscience, was performed in order to measure the root mean square roughness (S_q) of the samples in averaged over an area of 100 µm x 100 µm. Electrodeposition of Copper onto the plate surface was done in a solution containing 40 g/l of Copper Oxide and 160 g/l of H_2SO_4 at room temperature. Both plates, NiTi cathode and Lead anode, had an active surface of 1 cm2 for electrodeposition. The electrodeposition was performed under continuous stirring to reduce the diffusion layer. The current was controlled by a standard power source. Current intensity was kept constant at 200 A/m^2 for 20 minutes. Figure 1 depicts the experimental setup used to perform the electrodeposition. The plate mass weight was measured prior to and after electrodeposition in order to estimate the amount of deposited Copper.

IOP Conf. Series: Materials Science and Engineering **258** (2017) 012009 doi:10.1088/1757-899X/258/1/012009

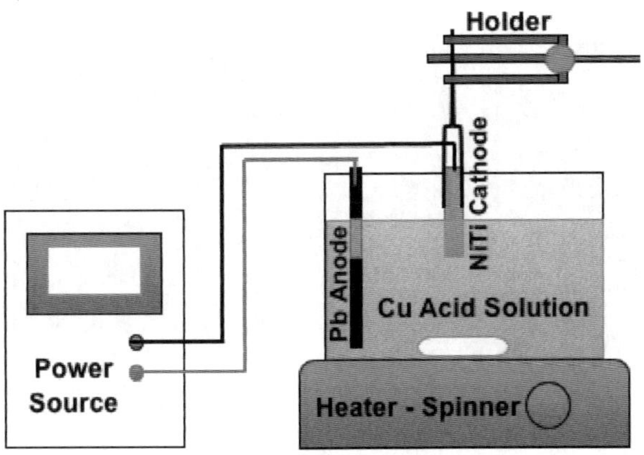

Figure 1. Schematic diagram of the experimental setup used to perform the electrodeposition of Copper on the NiTi substrates.

The characterization of the samples was performed using a laser-scanning microscope OLS40-SU (Olympus) and a dual beam scanning electron microscope (SEM) Helios NanoLab TM600 (FEI) equipped with focus ion beam (FIB) system. Grazing incidence X-ray diffraction (GI-XRD) was performed using the X-ray diffractometer MPD PANalyical X'Pert with Cu Kα1 radiation (λ = 0.15406 nm) generated at 40 kV and 40 mA. The primary optic was divergent (slit of 1°) and the X-ray source was fixed allowing an angle of incidence of ω = 4° between the X-rays and the sample surface. The secondary optic was divergent (slit of 1 mm) and the detector scanned between 2θ = 5° and 2θ = 140° with step size of 2θ = 0,04°. Differential scanning calorimetry (DSC) was conducted using the equipment Perkin-Elmer DSC-6000 at the scanning rate of 10 °C·min^{-1} in a rich N_2 atmosphere (99,95 % of purity).

3. Results and Discussion

3.1. Topography

Before the etching process on NiTi, the roughness S_q measured by AFM was 320 nm, whereas after the etching process an increase in S_q of 104 nm was observed. Therefore, an effective depth of around 50 nm produced by etching effect is estimated. Three Nitinol substrates were exposed to Copper deposition under the same experimental conditions. They were weighted prior to and after deposition, resulting in a Copper mass of 0.019 $\pm$ 0.006 g, which corresponds to 20 $\pm$ 6 μm of coating thickness. It is important to note that the calculated coating thickness is an average value due to differences in the distribution of electric field at the cathode surface. The samples were imaged using laser- and electron-scanning microscopy. Figure 2 shows optical images of the as-coated samples (inset) as well as top- and transversal-view images of the Copper coating electrodeposited onto Nitinol. As a standard procedure, Pt layers are deposited either by Ga-beam (Pt-Ga) and/or electron-beam (Pt-e) evaporation in the SEM microscope in order to avoid sample damage due to the impact of Ga ions during the FIB cut.

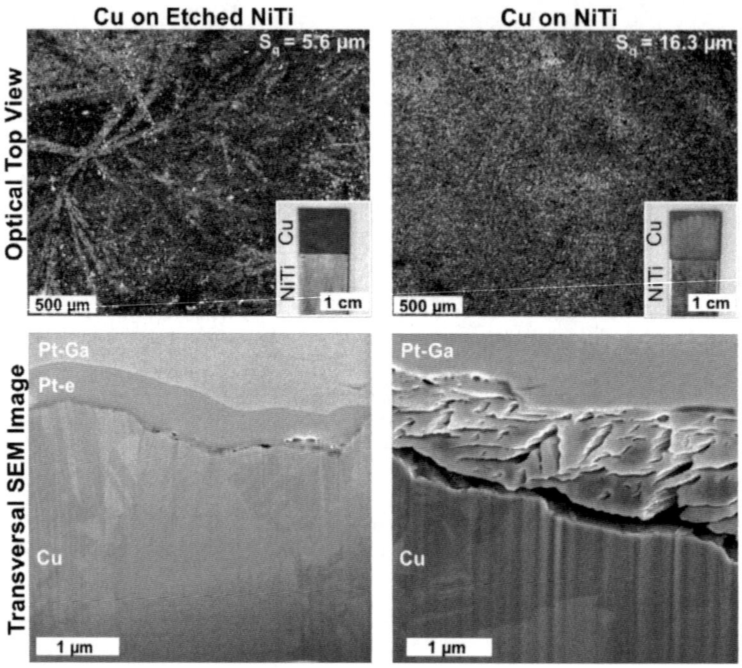

Figure 2. Top-view and transversal images of the Copper coating electrodeposited on Nitinol with and without etching pre-treatment. Pt layers are deposited prior to the FIB cut in order to avoid sample damage. Inset: optical micrographs of the as-coated samples showing the NiTi substrate and the Copper coating.

Laser-scanning microscopy (top images in Figure 2) show the presence of 2D dendritic-like structures at the surface of the Copper coating deposited on the Nitinol subjected to etching. These structures cannot be clearly observed for the other sample without etching, although both images were obtained under the same optical conditions. Dendritic structures in deposition of Copper have being observed for high voltages under current saturation [16-18]. Dendritic shapes in metals are related with the nucleation and growth kinetics in which free metal atoms continuously minimize the growth area by maximizing the surface energy.

The S_q parameter determined by Laser-scanning microscopy was found to be 5.6 µm for the etched Nitinol sample and 16.3 µm for the non-etched Nitinol sample. Moreover, SEM/FIB cross-sections (bottom images in Figure 2) show defined Cu grains underneath the surface in both samples and a porous top layer of around 2 µm in the non-etched Nitinol sample. This layer might correspond to a discontinuity of the growth kinetics due to fluctuations of the deposition process such as excessive turbulences and/or local temperature variations in the electrodeposition solution [17,18].

3.2. Chemical Analysis

In order to study the species formed at the NiTi-Copper interface, GI-XRD was performed on both samples. All diffractograms were obtained using the same experimental conditions (optic, measuring time, step size and so on) by keeping the X-ray source at a fixed angle of 4° with respect to the sample surface and moving the detector to perform the 2θ-scan. With this geometry maximum information depth was approximately 4 µm. Figure 3 shows diffractograms with the respective peak assignment for both samples. Both samples show rather small NiTi peaks (PDF 35-1281) due to the small information depth and the absorption of the X-rays in the coating layer. In addition to that, both samples show strong Cu peaks (PDF 4-836), whereas only the etched sample shows several peaks that correspond to the $CuSO_4(5H_2O)$ (PDF 77-1900) due to reduction reactions. The reduction may have happened at the surface of the NiTi, which was directly affected by the etching process.

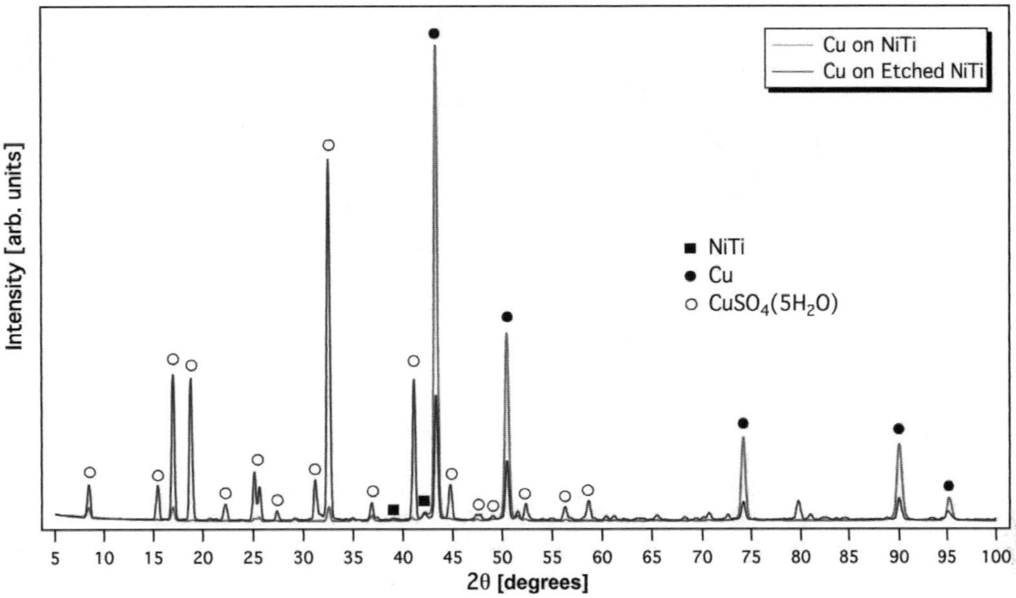

Figure 3. GI-XRD patterns of the Copper coated NiTi samples with and without etching prior to electrodeposition.

3.3. Thermal Analysis

In order to study the effect of the etching process and the electrodeposition of Copper on the thermal transformations of Nitinol, DSC was performed on three samples, namely: reference NiTi, etched NiTi and etched-coated NiTi. Figure 4 shows the endothermic (heating) and exothermic (cooling) behavior of those samples for two consecutive cycles in order to separate the information of the sample bulk from the adsorbents. The samples weight for the DSC analyses were 18.5 mg (Reference NiTi), 19.0 mg (Etched NiTi) and 36.0 mg (Etched and Cu-coated NiTi).

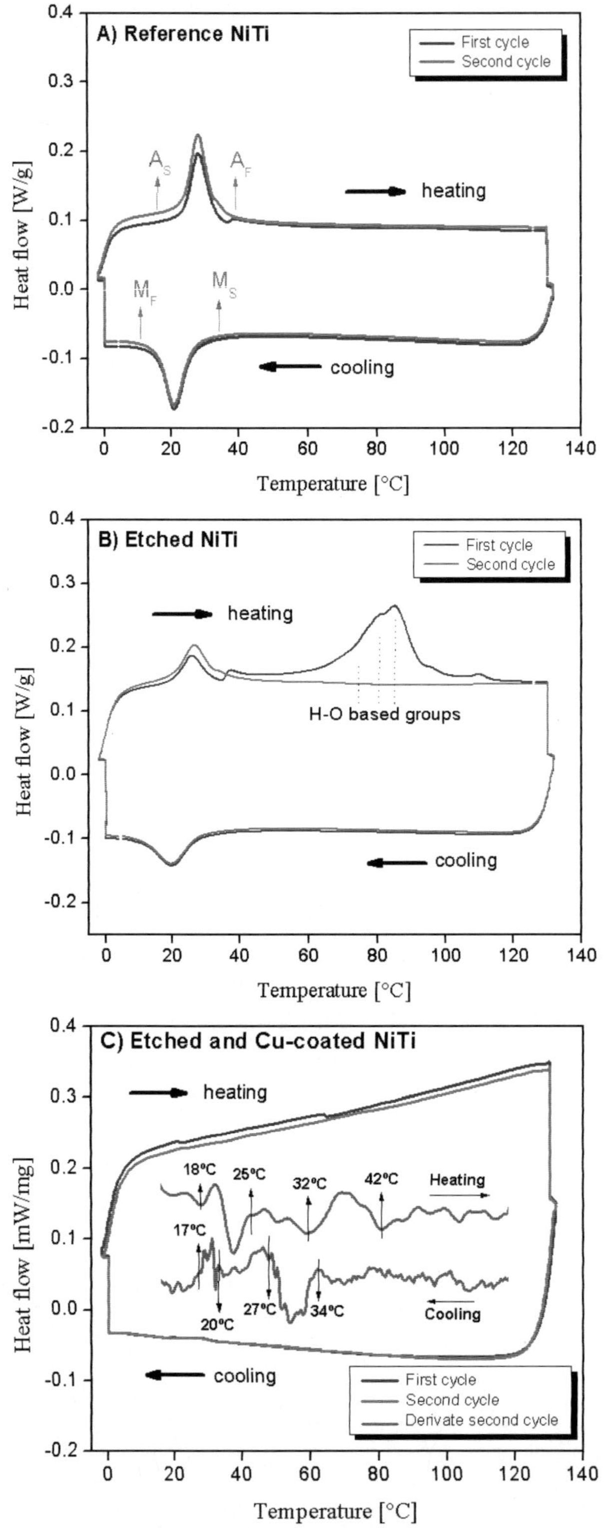

Figure 4. DSC curves measured on NiTi reference sample (A), NiTi after etching treatment (B), and NiTi after etching and Copper electrodeposition (C).

IOP Conf. Series: Materials Science and Engineering **258** (2017) 012009 doi:10.1088/1757-899X/258/1/012009

The reference and etched NiTi samples show the endothermic austenitic and exothermic martensitic phase transformation at similar temperatures below 40 °C. This is a well known behavior of NiTi and is associated with shape memory properties [2,19]. It is worth to note that the first cycle of the etched sample shows intense endothermic signals above 60 °C, which corresponds to surface diffusion and/or desorption of O-H based groups, including water.

In order to study the phase transformations, three parameters were analyzed: the starting transformation temperature (A_S and M_S), the final transformation temperature (A_F and M_F) and the enthalpy (ΔH_A and ΔH_M). Table 1 resumes those parameters for the austenite and martensitic transformations. These temperatures correspond to the points where the curve crosses the straight baseline, whereas the enthalpies were estimated by integrating the DSC curves between both transformation temperatures. Typically, transformation from martensite to austenite undergoes a specific enthalpy change of around 15 J/g [20], but this value is very sensitive to changes in composition [21] and grain boundaries density [22].

Table 1: Temperatures and enthalpies associated with the austenite and martensitic transformations, for all the analyzed samples.

	Austenitic Transition			Martensitic Transition		
Sample	AS [°C]	AF [°C]	ΔHA [J/g]	MS [°C]	MF [°C]	ΔHM [J/g]
Reference NiTi	17	40	4.6	31	11	3.9
Etched NiTi	17	38	2.5	33	6	3.3

The etched-coated NiTi sample (Figure 4c) shows linear heating and cooling behavior due to the presence of Cooper and no adsorbed H-O groups was observed. The linear behavior corresponds to materials with low heat capacities that experience high variation of temperature with low amounts of heat. In this case, although the specific heat of the Copper and Nitinol are similar [23], the Copper coating has a very low comparative mass and therefore behaves as a material with very low thermal capacity. Nevertheless, after derivation of the second DSC cycle, two small inflections can be observed for each endothermic and exothermic range. As shown in Figure 4, the endothermic inflections are observed between 18 and 25 °C as well as between 32 and 42 °C, whereas the exothermic inflections are observed in the intervals of 34 to 27 °C and 20 to 17 °C.

The analysis of the DSC data (see Table 1) shows that etching on Nitinol produced a reduction of 46 and 15% in the austenitic and martensitic transformation enthalpies, respectively. It is well known that changes in enthalpies might be produced by a lost of thermo-elastic energy after the generation of twins, impurities, defects and/or dislocations [19,20,24]. Moreover, the content of Cu in Ni-Ti-Cu alloys is known to directly affects the shape memory properties [21,25], suggesting and effect of Copper diffusion into Nitinol. We claim that etching consequently affects the surface of the Nitinol by generating energy sinks that modify the amount of energy driven to the shape memory effect.

Although our observations are interesting, they cannot be tracked back directly to phase transitions and therefore will be studied in more detail in a follow-up manuscript including mechanical testing such as stress-strain analyses.

4. Conclusions

1. Surface etching and electrodeposition of Copper coatings was successfully achieved on Nitinol substrates with shape memory properties.
2. Surface etching on Nitinol induces the formation of $CuSO_4(5H_2O)$ at the Cooper-Nitinol interface and reduces the surface roughness of electrodeposited Copper coating.
3. A reduction of austenitic-martensitic transformations enthalpies were observed after surface etching on Nitinol, suggesting an influence on the shape memory effect.
4. Further studies are needed in order to highlight the influence of the electrodeposited Copper on the memory shape properties of NiTi.

5. Acknowledgements

E. Ramos-Moore greatly acknowledges Sebastián Suárez and the Chair of Functional Materials at Saarland University for SEM/FIB characterization. E. Ramos-Moore also thanks the financial support of the FONDECYT 1150357 grant project funded by CONICYT Chile. J. Ramos-Grez acknowledges Fondecyt 1130906 grant project funded by CONICYT Chile. A. Rosenkranz greatly acknowledges the financial support of CIEN UC (Centro de Investigación en Nanotecnología y Materiales Avanzados) for his postdoctoral fellowhsip.

References

[1] A. Ölander, An electrochemical investigation of solid cadmium-gold alloys, J. Am. Chem. Soc. (1932).

[2] J.M. Jani, M. Leary, A. Subic, M.A. Gibson, A review of shape memory alloy research, applications and opportunities, Materials and Design. 56 (2014) 1078–1113. doi:10.1016/j.matdes.2013.11.084.

[3] W. Cai, X.L. Meng, L.C. Zhao, Recent development of TiNi-based shape memory alloys, Current Opinion in Solid State and Materials Science. 9 (2005) 296–302. doi:10.1016/j.cossms.2006.07.002.

[4] W. Cai, X.L. Lu, L.C. Zhao, Damping behavior of TiNi-based shape memory alloys, Materials Science and Engineering: A. 394 (2005) 78–82. doi:10.1016/j.msea.2004.11.030.

[5] X.T. Zu, C.F. Zhang, S. Zhu, Y. Huo, Z.G. Wang, L.M. Wang, Electron irradiation-induced changes of martensitic transformation characteristics in a TiNiCu shape memory alloy, Materials Letters. 57 (2003) 2099–2103. doi:10.1016/S0167-577X(02)01145-X.

[6] S. Shabalovskaya, J. Anderegg, J. Van Humbeeck, Critical overview of Nitinol surfaces and their modifications for medical applications, Acta Biomaterialia. 4 (2008) 447–467. doi:10.1016/j.actbio.2008.01.013.

[7] F. El Feninat, G. Laroche, M. Fiset, D. Mantovani, Shape memory materials for biomedical applications, Adv. Eng. Mater. 4 (2002) 91–104. doi:10.1002/1527-2648(200203)4:3<91::AID-ADEM91>3.0.CO;2-B.

[8] W. Theisen, A. Schuermann, Electro discharge machining of nickel-titanium shape memory alloys, Materials Science and Engineering: A. 378 (2004) 200–204. doi:10.1016/j.msea.2003.09.115.

[9] D.A. Armitage, T.L. Parker, D.M. Grant, Biocompatibility and hemocompatibility of surface-modified NiTi alloys, Journal of Biomedical Materials Research Part A. 66A (2003) 129–137. doi:10.1002/jbm.a.10549.

[10] X. Kong, R.G. Grabitz, W. Van Oeveren, D. Klee, T.G. van Kooten, F. Freudenthal, et al., Effect of biologically active coating on biocompatibility of Nitinol devices designed for the closure of intra-atrial communications, Biomaterials. 23 (2002) 1775–1783. doi:10.1016/S0142-9612(01)00304-0.

[11] R.A. Ahmed, S.A. Fadl-allah, N. El-Bagoury, S.M.F.G. El-Rab, Improvement of corrosion resistance and antibacterial effect of NiTi orthopedic materials by chitosan and gold nanoparticles, Applied Surface Science. 292 (2014) 390–399.

doi:10.1016/j.apsusc.2013.11.150.

[12] Y.F. Zheng, B.B. Zhang, B.L. Wang, Y.B. Wang, L. Li, Q.B. Yang, et al., Introduction of antibacterial function into biomedical TiNi shape memory alloy by the addition of element Ag, Acta Biomaterialia. 7 (2011) 2758–2767. doi:10.1016/j.actbio.2011.02.010.

[13] G. Reyes Donoso, M. Walczak, E.R. Moore, J.A. Ramos-Grez, Towards direct metal laser fabrication of Cu-based shape memory alloys, Rapid Prototyping Journal. 23 (2017) 329–336. doi:10.1108/RPJ-02-2016-0017.

[14] A. Ishida, M. Sato, Z.Y. Gao, Properties and applications of Ti-Ni-Cu shape-memory-alloy thin films, Journal of Alloys and Compounds. 577 (2013) S184–S189. doi:10.1016/j.jallcom.2011.12.155.

[15] N. Ucar, N. Turku, A.F. Ozdemir, A. Calik, Boriding of Binary Ni-Ti and Ternary Ni-Ti-Cu Shape Memory Alloys, Acta Physica Polonica A. 130 (2016) 492–495. doi:10.12693/APhysPolA.130.492.

[16] R. Qiu, H.G. Cha, H.B. Noh, Y.B. Shim, X.L. Zhang, R. Qiao, et al., Preparation of Dendritic Copper Nanostructures and Their Characterization for Electroreduction, J. Phys. Chem. 113 (2009) 15891–15896. doi:10.1021/jp904222b.

[17] N. Nikolic, K. Popov, L. Pavlovic, M. Pavlovic, Determination of Critical Conditions for the Formation of Electrodeposited Copper Structures Suitable for Electrodes in Electrochemical Devices, Sensors 2007, Vol. 7, Pages 1-15. 7 (2007) 1–15. doi:10.3390/s7010001.

[18] N.D. Nikolić, K.I. Popov, L.J. Pavlović, M.G. Pavlović, Morphologies of copper deposits obtained by the electrodeposition at high overpotentials, Surface & Coatings Technology. 201 (2006) 560–566. doi:10.1016/j.surfcoat.2005.12.004.

[19] J. Uchil, K.K. Mahesh, K. Ganesh Kumara, Calorimetric study of the effect of linear strain on the shape memory properties of Nitinol, Physica B: Condensed Matter. 305 (2001) 1–9. doi:10.1016/S0921-4526(01)00593-2.

[20] G.B. Stachowiak, P.G. McCormick, Shape memory behaviour associated with the R and martensitic transformations in a NiTi alloy, Acta Metallurgica. 36 (1988) 291–297. doi:10.1016/0001-6160(88)90006-5.

[21] O. Mercier, K.N. Melton, The substitution of Cu for Ni in NiTi shape memory alloys | SpringerLink, Mta. (1979).

[22] A. Sepulveda, R. Muñoz, F. Lovey, C. Auguet, A. Isalgue, V. Torra, Metastable effects on martensitic transformation in SMA, Journal of Thermal Analysis and Calorimetry. 89 (2006) 101–107. doi:10.1007/s10973-005-7480-3.

[23] Q.C. Tian, J.S. Wu, Characteristic of specific heat capacity of NiTi alloy phases, Transactions of Nonferrous Metals Society of China. 10 (2000) 737–740.

[24] S.W. Robertson, A.R. Pelton, R.O. Ritchie, Mechanical fatigue and fracture of Nitinol, International Materials Reviews. 57 (2013) 1–37. doi:10.1179/1743280411Y.0000000009.

[25] Y. Xu, X. Huang, A.G. Ramirez, Crystallization of amorphous NiTiCu thin films, Journal of Alloys and Compounds. 480 (2009) L13–L16. doi:10.1016/j.jallcom.2009.01.113.